NUMERICAL
METHODS

PRENTICE-HALL INTERNATIONAL SERIES IN ENGINEERING

BLACKWELL AND KOTZEBUE *Semiconductor-Diode Parametric Amplifiers*
DRESHER *Games of Strategy: Theory and Application*
EDMUNDSON, ED. *Proceedings of the National Symposium on Machine Translation*
KOLK *Modern Flight Dynamics*
STANTON *Numerical Methods for Science and Engineering*

PRENTICE-HALL, INC.

PRENTICE-HALL INTERNATIONAL, INC., UNITED KINGDOM AND EIRE

PRENTICE-HALL OF CANADA, LTD., CANADA

Ralph G. Stanton

Professor of Mathematics
and Chairman of the Department
University of Waterloo
Waterloo, Ontario, Canada

NUMERICAL METHODS FOR
SCIENCE
AND ENGINEERING

Prentice-Hall, Inc.

Englewood Cliffs, New Jersey

PRENTICE-HALL INTERNATIONAL SERIES IN APPLIED MATHEMATICS

Current printing (last digit):

15   14   13   12   11   10   9   8   7   6

Library of Congress Catalog Card Number: 61-6422

PRINTED IN THE UNITED STATES OF AMERICA

62662–C

Preface

At the present time, numerical methods are assuming a role of increasing importance in applied science; the actual, rather than the theoretical production of a solution is of paramount interest, and more and more engineering curricula are including courses in numerical methods. With this in mind, the Engineering Physics curriculum at the University of Toronto was revised, several years ago, to include a basic course in numerical methods in the sophomore year; the present volume is an outgrowth of the course of lectures which I was asked to organize at that time. The students who originally took the course had the usual background in algebra, trigonometry, plane and solid geometry, and a fairly intensive manipulative course in calculus; while taking numerical methods, they were also taking two courses in analysis—one in differential calculus, the other in integral calculus and differential equations. Training roughly equivalent to the courses just mentioned should suffice as a background for working through this book.

At present, there are a number of books on numerical methods available; the majority of these are at a more advanced level than this volume. I have found that most of them are unsuitable as elementary texts on two grounds: first, they are not written as texts, but as advanced reference books which presuppose a great deal of prior knowledge on the part of the reader; secondly, they attempt to cover a great many topics, and end up by being far too voluminous for an elementary course. I have tried in this book to write a basic text that can be useful to undergraduates who have a good knowledge of calculus and differential equations; I have definitely not aimed at a book suitable as a reference work for advanced and graduate students—there are already plenty of excellent books of that type.

In writing such an elementary text, I have inevitably been confronted by the problem of selection: Should I include the Chebyshev polynomials?

v

How much work on matrices should be included? Is the Choleski method of matrix inversion a desirable topic? In all cases I have tried to be guided, not by a desire to give an exhaustive (or exhausting) treatment, but by the question "Is this a basic topic, or is it subsidiary, something in the nature of a secondary frill which the student can learn for himself when and if he needs it?" Naturally, not every one will agree with me about all the topics excluded; I shall be content if it is felt that the topics included do provide a good groundwork, of intrinsic value to the student in his training, and one from which he can easily proceed to further study.

To exemplify the process of selection referred to in the previous paragraph, I might mention two topics. Under matrices, I have stressed the operational approach; I feel that it is much more important for the beginning student to become familiar with a matrix as an entity, so that he can pass immediately and comfortably from the system of equations $AX = C$ to the solution $X = A^{-1}C$, than it is for him to learn innumerable elegant numerical methods of working with matrices; the elegancies can come at a later stage—indeed, they may never need to come—but the groundwork of matric notation must be laid at an early stage. Again, one of the most important applications of relaxation procedures is to the solution of partial differential equations; yet I have given only one numerical example of this precedure. On the other hand, throughout the book great emphasis has been laid on the method of relaxation in its basic form, as applied to ordinary simultaneous equations, and on the idea of iteration—of a step-wise approach to the solution of any problem. It is my experience that students, even when they have not previously studied the numerical solution of partial differential equations, have little trouble in tackling an equation such as the Poisson equation, provided that they do have a thorough understanding of the ideas of relaxation and iteration. They can solve partial differential equations by employing their previous knowledge of these two basic techniques, together with some reading on the topic of partial differential equations. More generally, I would say that an introductory course should have as much fat as possible stripped from it; if the student can see the skeletal outlines of the course, he himself can later clothe them with the interesting but less essential details.

It is inevitable that there should be a certain amount of disagreement as to what topics are most essential, or even most useful; for instance, one person will say that the Lagrange Formula (2.7) is "extensively used in the case of equal intervals for desk-calculation", whereas another will "admit that his experience has not made him partial to blind Lagrangian interpolation, except when special circumstances point very definitely to it". Probably the main areas of disagreement will lie in my treatment of difference equations, partial differential equations, and electronic computers, in Chapters 9, 10, and 11, and in the considerable stress I have laid upon the numerous examples. In this connection, I should point out that Chapters 9 and 10 are merely brief introductions to the basic ideas of difference equations and partial differential equations; an exhaustive treatment of these topics could not be given in a short space in an elementary text. Chapter 11 is a general discussion of computers, and could be read at (or near) the beginning of the course. Finally, the stress on examples is due to my personal conviction that nothing is more fruitful in developing a student's knowledge of numerical methods than the study of special cases; even much research has had to begin in this way.

Although the course embodied in the book was developed from the standpoint of hand and desk-calculator techniques, it can be considered either as a study in its own right or as a first step in preparation for work with digital computers. It is my personal belief that there should be a natural progression from hand calculation to desk-calculators and then to electronic computers, according to the complexity of the problems. I can not sympathize with the view that students should start out with their eyes on an electronic computer—that they should run before they creep. Rather, I feel that the majority of workers in science and engineering can make great use of numerical methods without perhaps ever encountering a problem of sufficient length or complexity to justify programming it for an electronic computer. Even if the problem must ultimately be put on a computer, preliminary hand computation is always of value; it gives you a feeling about the problem, and about how large the numbers involved are. Such information is always of value since, despite the popularized nomenclature of "giant brains", computers are best described as the most unintelligent of slaves.

The problems included are an essential part of the course; the majority should be worked, with exceptions in those cases in which several problems of one type have been included in order to provide a variety of choice. In most of the problems, hand techniques suffice; in about 10 per cent of them, a desk-calculator is almost essential to avoid excessive work; in another 10 per cent, it does save some work. However, I have twice given the course without even the assistance of a manually operated calculator; to do so merely involves easing up on the decimal requirements for a few of the solutions, say, dropping from six decimals to three, or the like.

Two final points must be made. The first is that I have tended to treat numerical methods as a part of empirical science rather than as a branch of pure mathematics. I am not interested in a puristic approach but in the fact that numerical methods give actual results which can be used in practical problems. The book is definitely intended for students in science and engineering; aesthetics must take second place to usability. In short, the book is addressed to an audience whose concern is with solving problems, not with talking about solving them.

The second point follows naturally from the first; operators have been used a great deal. Ever since the time of Heaviside, pure mathematicians have been worrying about the lack of rigour (and rigidity) involved in the use of operators. The fact remains that results can be obtained by using operators, and the development can be validated at a slightly higher mathematical level. Science and engineering students have to learn to use operators, and the emphasis on form and method of pure mathematics must yield to an emphasis on use and practicality in actual problems.

I would be ungrateful if I concluded this preface without expressing thanks to my former students in the Engineering Physics course at the University of Toronto, especially to that one who supplied me with the epigram "Numerical analysis is like engineering, partly a science, partly an art".

RALPH G. STANTON

Contents

ix

CHAPTER **11**

THE PRINCIPLES OF AUTOMATIC COMPUTATION

NUMERICAL
METHODS

Foreword to the
Student

Next to a pencil, the most important piece of equipment in numerical analysis is an eraser; however, assuming that these two necessities are in good supply, what accessories should the student possess?

A slide rule is useful; in many problems, the result is obtained by an iterative procedure which adds a small correction term at each stage of the process; if not more than three figures are required in this correction term, then the slide rule can save much work.

A good set of tables is indispensable; for student use, I have found that the Standard Mathematical Tables, published by the Chemical Rubber Publishing Company, are an invaluable and versatile set of tables. They are easily portable, contain a wealth of material, and should serve the student adequately throughout his university career.

Quadruled paper, available at almost any bookstore, is of the greatest assistance; the squares (quarter-inch) are of a convenient size to contain a single digit, and immeasurably reduce the problems of recording and utilizing computations.

Some students might like to invest in a small portable manually-operated desk calculator. Of those available, my personal preferences, on grounds of price, utility, and ease of understanding, lean towards the hand-operated Facit and the Monroe Educator model. More elaborate models are likely to obscure those fundamental principles which the beginning student should observe in his personal calculator.

With regard to computations, students are always prone to ask "How many figures should I keep?" One authority has answered "As many as the machine will hold, or as are conveniently available". With this view, I largely concur; common sense must always decide, and it is impossible to give precise general rules. One has to look ahead towards what one desires in the way of an answer; then one keeps enough figures to produce the type of answer required.

The student should always be alive to the possibilities of short cuts, many of which can only be learned by discovery and practice; for example, any

1

student past the complete amateur stage would compute 34.7/99 by writing

$$\frac{34.7}{99} = \frac{34.7}{100}(1 + 10^{-2} + 10^{-4} + ...)$$

$$= .347 + .00347 + .0000347 + ...$$

$$= .350505.$$

Innumerable opportunities for techniques of this nature are continually arising. Occasionally, in hand work, the "short methods" of multiplication and division are useful (however, they do contain undesirable features, and should be employed with caution). In case the student has not met them, they are exemplified below.

Ordinary Multiplication	*Short Method*
34.273	34273
16.121	12161
34273	34273
68546	20564
34273	343
205638	69
34273	3
552.515033	552.52

In the short method of multiplication, the following points may be noted:
1) the multiplier is reversed;
2) any digit in the multiplier acts first on the digit of the multiplicand which occurs directly above it (but we must not forget to make a mental carry-over from the unused digits on the right);
3) the decimal is located by inspection, as with the slide rule.

Ordinary Division	*Short Method*
5.3549	5.3549
65.217 / 349.23	65.217 / 349.23
326.085	326.09
23.1450	23.14
19.5651	19.56
3.57990	3.58
3.26085	3.26
319050	32
260868	26
581820	6
586953	6

In the short method of division, we note:
1) at each stage we delete one digit of the divisor rather than add an extra zero to the dividend;
2) carrying figures are added on mentally;
3) the decimal point is located by inspection.

A few remarks concerning errors may be useful (computational errors, that is, since experimental errors fall within the province of statistics). First, a caution: *the way in which a number is written tells nothing whatsoever about the number*; thus, if 3.1416 is written down, there is no way of knowing whether 3.1416 is standing in its own right or whether it merely represents a five-figure approximation to π; *information of such a nature comes from the particular problem being done, and not from the number itself*. For this reason, it is best to avoid the term "significant figure"; it is seldom applicable.

In this connection, we must say a few words about computations involving numbers which are obtained by measurement. Suppose, for example, that we have a formula $P = ab$, and that the values $a = 7.6$, $b = 4.8$, are obtained by measurement. It has become very popular in some quarters to say "a and b are 'approximate numbers' each given to two significant figures; so the product P is an 'approximate number' with two significant figures, namely, 36". (Just how the integer 36 can be an "approximate number" is not explained.) Now let us analyze this balderdash according to the claims of its own proponents. According to them, 7.6 has two significant figures, that is, 7.6 means "some number between 7.55 and 7.65"; similarly, 4.8 means "some number between 4.75 and 4.85", and 36 means "some number between 35.5 and 36.5". However, an elementary computation establishes the fact that ab, understood in their sense, lies between 35.8625 and 37.1025; this range bears little resemblance to the range 35.5–36.5 which they give; indeed, the mean value 36.48 for the range barely lies within the range given by their own answer. Thus the statement $7.6 \times 4.8 = 36$ is completely ridiculous, even using the interpretation attached to it by its own proponents; for this reason, the contradictory and misleading term "approximate number" should be avoided like the plague; it leads to ludicrous results. The only sensible way of approaching the above problem is to write: $a = 7.6$, $b = 4.8$; then $P = 36.48$. One understands that if there are errors δ_1 and δ_2 associated with the values a and b respectively, then there will be an error ϵ associated with the computed value of P; ϵ will be a function of δ_1 and δ_2, which are generally not known. Any statistical discussion will require the estimate 36.48 of P, not an estimate of this estimate.

Computational errors should either be errors of truncation (due to cutting off infinite formulae at a finite stage) or errors of round-off (due to cutting off infinite decimals at a finite stage). Truncation errors have to be considered on their own merits; round-off errors are always at most half a unit in the last figure recorded, but they may accumulate, in the course of a problem, by the composition of several round-off errors. For this reason, it is necessary to maintain one or more guard figures in computations; they will not, in general, be quite correct themselves, but they are needed to protect the correctness of earlier figures. Thus, if a final answer correct to five figures is required, it will be necessary to keep six or seven figures in intermediate calculations; rounding to the required number of figures should *never* be done before the final answer has been reached.

It might be noted that mistakes or blunders are not included under the heading of errors. Mistakes can be checked, corrected, and avoided; error accumulation is unavoidable because of the inability of either man or machine to record an infinite decimal in any other way than by replacing it by a differing finite decimal.

In conclusion, we must say a final word concerning numbers obtained as representing measurements and their relation to hypothetical "true values". The commonest incorrect statement made in this connection is to the effect that "the possible error a trained observer may make will not exceed one-half the unit which corresponds to the last significant digit in the measure (*sic*)". Thus a measurement of 14.237 is said to indicate a true value lying between 14.2365 and 14.2375. That this is an unwarranted degree of optimism can be readily illustrated by an example. Crookes made ten determinations of the atomic weight of thallium (*Phil. Trans.*, **163**, p. 277, 1874), and gave his results as 203.628, 203.632, 203.636, 203.638, 203.639, 203.642, 203.644, 203.649, 203.650, 203.666. Now Crookes was certainly a trained observer, and if we discard one decimal place, all his values lie between 203.63 and 203.67; yet if we consult a modern list of atomic weights, we find that of thallium given as 204.39. The history of science is full of other illustrations of the fact that true values may differ greatly from estimates of them based on measurements, even when those measurements have been made by "trained observers". Indeed, the dictum quoted at the beginning of this paragraph betokens a complete confusion between the concepts of numbers obtained by rounding off and numbers obtained as representing measurements; there is no similarity whatever between the two concepts. If we round off $\sqrt{2}$ to the value 1.414, we know that the true value of $\sqrt{2}$ lies between 1.4135 and 1.4145; but if we measure a quantity as 1.414, it would indeed be rare that we were so enamoured of our determination as to claim that the true value was in such a small range as 1.4135–1.4145. The treatment of experimental data falls properly within the scope of statistical methods, and need not concern us at all in the present volume; however, we do repeat the caution: *Numbers representing measurements must not be treated as if they resulted from rounding off.* In this connection, the student might very profitably read the excellent and stimulating article "Computations with Approximate Numbers" by D. B. De Lury in *The Mathematics Teacher*, November, 1958.

For the convenience of the student, a list of the formulae required in this book follows this foreword.

Review of Formulae

1. *Sigma Notation.*

$$\sum_{i=0}^{n} a_i = a_0 + a_1 + \ldots + a_n,$$

$$\sum_{i=r}^{\infty} a_i = a_r + a_{r+1} + \ldots + a_n + \ldots,$$

$$\sum_{i=0}^{n} a_i x^i = a_0 + a_1 x + a_2 x^2 + \ldots + a_n x^n$$

(i, r, n, integers).

2. *Pi Notation.*

$$\prod_{i=1}^{n} a_i = a_1 a_2 a_3 \ldots a_n.$$

3. *Factorial Notation.*

$$r! = r(r-1)(r-2) \ldots 3 \cdot 2 \cdot 1 = \prod_{x=1}^{r} x.$$

$$0! = 1.$$

4. *The Binomial Coefficients.*

$$\binom{n}{r} = \frac{n(n-1)(n-2) \ldots (n-r+1)}{r!}$$

(r a non-negative integer; n arbitrary).

$$\binom{n}{r} = {}_nC_r \quad \text{if } n \text{ is a positive integer.}$$

5. *The Binomial Theorem.*

$$(a+x)^n = \sum_{i=0}^{n} \binom{n}{i} x^i a^{n-i} \quad (n \text{ a positive integer}).$$

$$(1+x)^n = \sum_{i=0}^{\infty} \binom{n}{i} x^i$$

(n arbitrary; this series is numerically meaningful for $|x| < 1$).

6. *Binomial Identities.*

$$\binom{n}{0} = 1; \qquad\qquad \binom{n}{r} = 0 \quad \text{for} \quad r < 0.$$

$$\binom{n}{r} = 0 \qquad\qquad (n \text{ a positive integer}, r > n).$$

$$\binom{n}{r} = \binom{n}{n-r} \qquad (n \text{ and } r \text{ positive integers}).$$

$$\binom{n-1}{r} + \binom{n-1}{r-1} = \binom{n}{r}.$$

7. *Mathematical Induction* (*n* integral).

Let $P(n)$ be a statement such that
  (i) $P(a)$ is true;
  (ii) the truth of $P(n)$ implies the truth of $P(n + 1)$.

Then $P(n)$ is true for all integers $n > a$. In applications, a is usually 0 or 1.

8. *Interval Notation.*

"From .1 (.02) .3" is to be interpreted as "from .1 to .3 at intervals of .02 units, that is, at the values .10, .12, .14, ... , .28, .30".

9. *The Kronecker Delta.*

$$\delta_{ij} = 1 \quad \text{if} \quad i = j, \qquad \delta_{ij} = 0 \quad \text{if} \quad i \neq j.$$

10. *Properties of the Roots of Polynomial Equations.*

Let

$$f(x) = x^n + a_{n-1}x^{n-1} + \ldots + a_2x^2 + a_1x + a_0 = 0$$

be a polynomial equation of degree n with real or complex coefficients a_i. Then $f(x) = 0$ has exactly n roots r_i, and we may write

$$f(x) = (x - r_1)(x - r_2) \ldots (x - r_n) = 0.$$

The r_i may be real or complex, and not all need be distinct. If all the coefficients a_i are real, then any complex roots of $f(x) = 0$ must occur in conjugate complex pairs of the form $a \pm bi$. Furthermore, the *elementary symmetric functions* of the r_i are given by the equations

$$\Sigma_1 \, r_i = r_1 + r_2 + \ldots + r_n = -a_{n-1},$$
$$\Sigma_2 \, r_i r_j = r_1 r_2 + r_1 r_3 + \ldots + r_{n-1} r_n = a_{n-2},$$
$$\Sigma_3 \, r_i r_j r_k = -a_{n-3},$$
$$\ldots\ldots\ldots\ldots\ldots\ldots\ldots\ldots$$
$$r_1 r_2 \ldots r_n = (-)^n a_0.$$

In all these formulae, the sigma sign Σ_u denotes a summation over all the $_nC_u$ possible products of the n roots taken u at a time.

11. *Approximation Sign.*

The sign "$\doteq$" is read as "is approximately equal to".

12. *Total Differential.*

If $f(u, v)$ is a function of two variables u and v, the total differential of f is given by the formula

$$df = \frac{\partial f}{\partial u}\, du + \frac{\partial f}{\partial v}\, dv.$$

If u and v undergo small increments du and dv, then df is the approximate increment in f.

CHAPTER ONE

Ordinary Finite Differences

1. BUILDING A DIFFERENCE TABLE

Suppose that a function $f(x)$ is given, and that a table is formed of the functional values $f(a), f(a + h), f(a + 2h), \ldots$, where the *independent variable* (or *argument*) x proceeds at equally-spaced intervals; the (constant) difference between two consecutive values of x is called the *interval of differencing* and will be denoted by the letter h. Then the *difference operator* Δ is defined by the equation

$$(1.1) \qquad \Delta f(x) = f(x + h) - f(x).$$

The quantity $\Delta f(x)$ is called the *first difference* of $f(x)$, and is itself a function of x; consequently, we can repeat the operation of differencing to obtain the *second difference* of $f(x)$ according to

$$(1.2) \qquad \Delta^2 f(x) = \Delta[\Delta f(x)] = \Delta f(x + h) - \Delta f(x).$$

In general, the nth difference of $f(x)$ will be defined by the recursion relation

$$(1.3) \qquad \Delta^n f(x) = \Delta[\Delta^{n-1} f(x)] = \Delta^{n-1} f(x + h) - \Delta^{n-1} f(x).$$

For example, let $f(x) = x^3 - 3x^2 + 5x + 7$; starting with $a = 0$ as initial value, the functional values for 0, 2, 4, 6, 8, are tabulated in Table 1.1.

In this example, the analytical formula for $\Delta f(x)$ is

$$\Delta f(x) = (x + 2)^3 - 3(x + 2)^2 + 5(x + 2) + 7 - (x^3 - 3x^2 + 5x + 7)$$
$$= 6x^2 + 6.$$

8

Similarly,

$$\Delta^2 f(x) = 6(x + 2)^2 + 6 - (6x^2 + 6) = 24x + 24,$$

and

$$\Delta^3 f(x) = 24(x + 2) + 24 - (24x + 24) = 48.$$

It is clear that $\Delta^4 f(x) = \Delta^5 f(x) = \ldots = 0$. Thus the property that $\Delta^3 f(x)$ is equal to a constant for all x may be used to extend the table of functional values as far as we please; this is done in Table 1.1 (below the line) to give $f(10)$.

Table 1.1.   DIFFERENCE TABLE FOR THE FUNCTION
$$x^3 - 3x^2 + 5x + 7$$

x	$f(x)$	$\Delta f(x)$	$\Delta^2 f(x)$	$\Delta^3 f(x)$
0	7			
		6		
2	13		24	
		30		48
4	43		72	
		102		48
6	145		120	
		222		48
8	367		168	
		390		
10	757			

From the definition (1.1), it follows that the operator Δ obeys the laws

(1.4) $$\Delta\{f(x) + g(x)\} = \Delta f(x) + \Delta g(x),$$

and (c being a constant)

(1.5) $$\Delta\{cf(x)\} = c\,\Delta f(x).$$

Employing these two relations, we show that the property exemplified in Table 1.1 is a special case of

THEOREM 1.1. If $f(x)$ is a polynomial of degree n,

$$f(x) = \sum_{i=0}^{n} a_i x^i,$$

then $\Delta^n f(x)$ is constant, and is equal to $a_n n!\, h^n$.

Proof. For $n = 1$, $f(x) = a_1 x + a_0$ and $\Delta f(x) = a_1 h$; so the theorem is true for $n = 1$. Assume now that the theorem is true for all degrees $1, 2, \ldots,$ $n - 1$, and consider the n-ic polynomial

$$f(x) = \sum_{i=0}^{n} a_i x^i;$$

applying (1.4) and (1.5), we find

$$\Delta^n f(x) = \sum_{i=0}^{n} a_i \, \Delta^n x^i.$$

For $i < n$, $\Delta^n x^i$ is the nth difference of a polynomial of degree less than n, and hence must vanish, by the induction hypothesis. Thus

$$\Delta^n f(x) = a_n \, \Delta^n x^n$$
$$= a_n \, \Delta^{n-1}(\Delta x^n)$$
$$= a_n \, \Delta^{n-1}\{(x+h)^n - x^n\}$$
$$= a_n \, \Delta^{n-1}\{nhx^{n-1} + g(x)\},$$

where $g(x)$ is a polynomial of degree less than $n-1$. Hence, applying the induction hypothesis again, we find

$$\Delta^n f(x) = a_n \, \Delta^{n-1}(nhx^{n-1}) = a_n(nh)(n-1)! \, h^{n-1} = a_n n! \, h^n.$$

This proves the theorem by induction.

In concluding this section, we might note that very often the argument x is denoted by a subscript; thus, instead of $f(x)$, we could write f_x. The advantage of this alternative notation is in avoiding excessive use of brackets and parentheses; for example, we could write (1.4) in the simpler-appearing form $\Delta(f_x + g_x) = \Delta f_x + \Delta g_x$.

EXERCISES

1. Form the difference table of

$$f_x = x^4 - 5x^3 + 6x^2 + x - 2$$

   for values $x = -3, -2, -1, 0, 1, 2, 3$; extend the table in both directions to give f_4, f_5, f_6, f_7.

2. With $h = 1$, give the analytical expressions for Δf_x, $\Delta^2 f_x$, $\Delta^3 f_x$, when $f_x = x^3 - 7x^2 + 2x + 3$.

3. Find the analytical expression for $\Delta^n(c^x)$, c being a constant.

4. The expression $x^{(r)} = x(x-h)(x-2h)\dots(x-\overline{r-1}\,h)$ occurs frequently; prove that $\Delta x^{(r)} = rhx^{(r-1)}$.

5. With $h = 1$, express $f_x = x^3 - 7x^2 + 2x + 2$ in the form

$$f_x = \sum_{i=0}^{3} a_i x^{(i)};$$

   thence obtain $\Delta f(x)$. (Note: by convention, $x^{(0)} = 1$.)

6. Find the next two terms of the sequence

$u_0 = 5, u_1 = 11, u_2 = 22, u_3 = 40, u_4 = 74, u_5 = 140, u_6 = 261, u_7 = 467,$

given that the general term u_n is represented by a quartic polynomial in n.

7. Using $x = 1, 2, \ldots , 8$ as values of the argument, form difference tables (as far as fourth differences) for the functions $f(x) = x^4$ and $g(x) = 3^x$. Will the differences of $g(x)$ ever become constant?

8. Find Δu_x, $\Delta^2 u_x$, $\Delta^3 u_x$ for the functions $u_x = ax^3 - bx + c$ and $u_x = 1/x$. Take $h = 1$.

9. Form a difference table (to fourth differences).

x	1	2	3	4	5	6	7	8
f_x	7.93	10.05	12.66	15.79	19.47	23.73	28.60	34.11

Repeat the procedure for the same table when $f_5 = 19.47 + \epsilon$. If ϵ represents an error in a single value f_x, how many entries $\Delta^n f_x$ are affected?

10. Find, from a difference table, the error made in recording the following values $u_0, u_1, \ldots , u_{12}$.

47.2, 49.3, 52.2, 56.2, 61.6, 68.7, 77.8, 89.2, 102.3, 120.1, 140.2, 163.8, 191.2.

11. Use mathematical induction to prove the binomial theorem

$$(1 + x)^n = \sum_{i=0}^{n} \binom{n}{i} x^i,$$

where n is a positive integer.

12. For n a positive integer, find an empirical formula for the sum

$$S_n = \frac{1}{1 \cdot 4} + \frac{1}{4 \cdot 7} + \frac{1}{7 \cdot 10} + \ldots + \frac{1}{(3n - 2)(3n + 1)},$$

and verify the formula by mathematical induction.

13. Use mathematical induction to prove that

$$1^3 + 2^3 + \ldots + n^3 = \frac{n^2(n + 1)^2}{4}.$$

2. THE OPERATORS E AND Δ

Consider the differences of the function f_x as given in Table 1.2.

We now introduce a second operator, the *enlargement operator* E (also called the displacement operator), defined by the equation

(1.6)　　　　　　　　　　　$E f_x = f_{x+h}.$

Thus we can think of E as the operator moving the functional value f_x along to the next higher value f_{x+h}; a second operation with E would give

$$E^2 f_x = E[Ef_x] = Ef_{x+h} = f_{x+2h},$$

and, in general,

(1.7) $$E^n f_x = E[E^{n-1} f_x] = Ef_{x+(n-1)h} = f_{x+nh}.$$

Consequently, in Table 1.2, we could write the functional values f_a, f_{a+h}, f_{a+2h}, ... in the alternative forms f_a, Ef_a, $E^2 f_a$, ...

Table 1.2. DIFFERENCES OF THE FUNCTION f_x

x	f_x	Δf_x	$\Delta^2 f_x$	$\Delta^3 f_x$	$\Delta^4 f_x$
a	f_a				
		Δf_a			
$a+h$	f_{a+h}		$\Delta^2 f_a$		
		Δf_{a+h}		$\Delta^3 f_a$	
$a+2h$	f_{a+2h}		$\Delta^2 f_{a+h}$		$\Delta^4 f_a$
		Δf_{a+2h}		$\Delta^3 f_{a+h}$	
$a+3h$	f_{a+3h}		$\Delta^2 f_{a+2h}$		
		Δf_{a+3h}			
$a+4h$	f_{a+4h}				

Our third operator will be the identity operator 1, with the property that it operates on f_x to leave f_x unaltered; in symbols,

(1.8) $$1f_x = f_x.$$

It can be related to the operators E and Δ by noting that

$$\Delta f_x = f_{x+h} - f_x = Ef_x - 1f_x,$$

or

(1.9) $$Ef_x = 1f_x + \Delta f_x.$$

This relationship (1.9) is often written in the form

(1.10) $$Ef_x = (1 + \Delta)f_x,$$

or simply

(1.11) $$E = 1 + \Delta.$$

It must be stressed that Equation (1.11) does not mean that the operators 1, E, and Δ have any existence as separate entities; it means exactly the same as Equation (1.10), that is, if the operator E acts on a function f_x, the result is the same as would be achieved by the action of the operator $1 + \Delta$.

It now becomes natural to try to express the higher differences in Table 1.2 in terms of the given functional values; we find experimentally

$$\Delta f_a = f_{a+h} - f_a,$$
$$\Delta^2 f_a = \Delta f_{a+h} - \Delta f_a = (f_{a+2h} - f_{a+h}) - (f_{a+h} - f_a)$$
$$= E^2 f_a - 2E f_a + f_a,$$
$$\Delta^3 f_a = \Delta^2 f_{a+h} - \Delta^2 f_a = (E^2 f_{a+h} - 2E f_{a+h} + f_{a+h}) - (E^2 f_a - 2E f_a + f_a)$$
$$= E^3 f_a - 3E^2 f_a + 3E f_a - f_a.$$

The preceding results certainly suggest a binomial relationship, especially when we notice that this is just what we would get if we treated 1, E, and Δ as formal mathematical quantities, and used (1.11) to give

$$\Delta^3 f_a = (E - 1)^3 f_a = (E^3 - 3E^2 + 3E - 1)f_a$$
$$= f_{a+3h} - 3f_{a+2h} + 3f_{a+h} - f_a.$$

Conversely, we should hope to be able to write

$$f_{a+3h} = E^3 f_a = (1 + \Delta)^3 f_a$$
$$= (1 + 3\Delta + 3\Delta^2 + \Delta^3)f_a$$
$$= f_a + 3\Delta f_a + 3\Delta^2 f_a + \Delta^3 f_a.$$

Indeed, if we could expand $1 + \Delta$ by the binomial theorem, treating Δ simply as a formal mathematical quantity, we should have

(1.12) $$f_{x+nh} = E^n f_x = (1 + \Delta)^n f_x = \sum_{i=0}^{\infty} \binom{n}{i} \Delta^i f_x.$$

We shall justify this formal expansion by giving an independent proof of Equation (1.12) in

THEOREM 1.2.

$$f_{x+nh} = \sum_{i=0}^{\infty} \binom{n}{i} \Delta^i f_x,$$

where n is a positive integer.

Proof. For $n = 1$, the theorem reduces to the statement $E f_x = f_x + \Delta f_x$, and this is just Equation (1.9). Assume now that the theorem is true for the value $n - 1$; then

$$E^n f_x = E[E^{n-1} f_x] = E \sum_{i=0}^{\infty} \binom{n-1}{i} \Delta^i f_x,$$

by the induction hypothesis. But, using (1.11), we get

$$E^n f_x = (1 + \Delta)E^{n-1}f_x = E^{n-1}f_x + \Delta E^{n-1}f_x$$

$$= \sum_{i=0}^{\infty} \binom{n-1}{i} \Delta^i f_x + \sum_{i=0}^{\infty} \binom{n-1}{i} \Delta^{i+1} f_x$$

$$= \sum_{i=0}^{\infty} \binom{n-1}{i} \Delta^i f_x + \sum_{j=1}^{\infty} \binom{n-1}{j-1} \Delta^j f_x.$$

The coefficient of $\Delta^k f_x$ ($k = 0, 1, 2, \ldots, n$) is given by

$$\binom{n-1}{k} + \binom{n-1}{k-1} = \binom{n}{k};$$

hence we conclude that

$$E^n f_x = \sum_{k=0}^{\infty} \binom{n}{k} \Delta^k f_x,$$

and this completes the proof of the theorem by induction.

The special case of (1.12) given by

(1.13) $$f_x = E^x f_0 = \sum_{i=0}^{\infty} \binom{x}{i} \Delta^i f_0$$

is usually called *Newton's Advancing Difference Formula*; it expresses the general functional value f_x in terms of f_0 and its differences.

We have proved (1.12) only in the case when n is a positive integer; however, provided that the higher differences decrease rapidly enough for the infinite series on the right-hand side of (1.12) to converge, it can be shown that (1.12) holds for all values of n. Indeed, for the purposes of our work, we shall frequently treat E, Δ, and other operators as if they were ordinary mathematical symbols capable of expansion by the binomial theorem, Taylor's series, or any other formula we wish to employ. In general, we shall omit proofs; the justification we have given for Equation (1.12) in the case that n is a positive integer is a sample of what can be done. We shall content ourselves with noting that "the formal use of operators works in practice and can be justified in theory".

EXAMPLE 1.1. Find the cubic polynomial u_x which takes on the values $u_0 = -5$, $u_1 = 1$, $u_2 = 9$, $u_3 = 25$, $u_4 = 55$, $u_5 = 105$. Use the difference table to compute $u_{3.2}$.

Solution. Forming the difference table for the function, we have

x	u_x	Δu_x	$\Delta^2 u_x$	$\Delta^3 u_x$
0	−5			
		6		
1	1		2	
		8		6
2	9		8	
		16		6
3	25		14	
		30		6
4	55		20	
		50		
5	105			

Then

$$u_x = E^x u_0 = (1 + \Delta)^x u_0$$

$$= \left[1 + x\Delta + \frac{x(x-1)}{2!}\Delta^2 + \frac{x(x-1)(x-2)}{3!}\Delta^3 \right] u_0$$

$$= u_0 + x\,\Delta u_0 + \frac{x^2 - x}{2}\Delta^2 u_0 + \frac{x^3 - 3x^2 + 2x}{6}\Delta^3 u_0.$$

Substituting $u_0 = -5$, $\Delta u_0 = 6$, $\Delta^2 u_0 = 2$, $\Delta^3 u_0 = 6$, we obtain the result

$$u_x = x^3 - 2x^2 + 7x - 5.$$

Interpolating to obtain $u_{3.2}$, we find

$$u_{3.2} = E^{.2}u_3 = (1 + \Delta)^{.2}u_3 = (1 + .2\Delta - .08\Delta^2 + .048\Delta^3)u_3$$

$$= 25 + .2(30) - .08(20) + .048(6) = 29.688.$$

EXERCISES

1. Form the difference table for the function

x	30	35	40	45	50	55
u	2.926	4.162	5.863	8.119	11.022	14.666

Write down the values of $\Delta^2 u_{30}$, Δu_{40}, $\Delta^3 u_{35}$, $\Delta^4 u_{35}$, $\Delta^5 u_{45}$. What values may be predicted from the table for u_{25}, u_{60}, and u_{42}?

2. Find, under a suitable assumption, the next two terms and the general term of the sequence

$$7, 9, 16, 32, 64, 122, 219, 371.$$

3. Find a function of x such that

$$u_1 = 3.2, \ u_3 = 9.6, \ u_4 = 28.6, \ u_5 = 20.5.$$

4. The following table gives the amount of a chemical dissolved in water.

Temperature	10	15	20	25	30	35
Solubility	19.97	21.51	22.47	23.52	24.65	25.89

Compute the amounts dissolved at $22°$, $8°$, $-15°$.

5. Compute $\Delta E f_x$ and $E\Delta f_x$; are the operators $E\Delta$ and ΔE the same?

6. Express in sigma notation and use (1.11) to prove the identity

$$u_1 + u_2 + \ldots + u_n = \binom{n}{1}u_1 + \binom{n}{2}\Delta u_1 + \ldots + \binom{n}{n}\Delta^{n-1}u_1.$$

7. Prove the identities

(a) $u_{x+n} = u_n + \binom{x}{1}\Delta u_{n-1} + \binom{x+1}{2}\Delta^2 u_{n-2} + \binom{x+2}{3}\Delta^3 u_{n-3} + \ldots ,$

(b) $\binom{n}{0}x - \binom{n}{1}(x-1) + \binom{n}{2}(x-2) - \binom{n}{3}(x-3) + \ldots = 0.$

8. Evaluate

$$(x-1)^2\binom{n}{1} + (x-3)^2\binom{n}{3} + (x-5)^2\binom{n}{5} + \ldots ,$$

and deduce that, for $x = n$, the sum reduces to $2^{n-3}n(n+1)$.

9. By writing

$$f(x) = x^6 - 27x^5 + 105x^4 - 140x^3 + 81x^2 - 21x + 2$$

in the form

$$f(x) = ([\{(x\overline{x-27} + 105)x - 140\}x + 81]x - 21)x + 2,$$

compute $f(x)$ for $x = -1, 0, 1, \ldots , 7$. Check your computation by finding Δ^6 from your table. Continue the table to obtain $f(8)$ and $f(9)$.

10. Find, under suitable assumptions, values for u_4 and u_6, given that $u_1 = 103.4$, $u_2 = 97.6$, $u_3 = 122.9$, $u_5 = 179.0$, $u_7 = 195.8$.

11. The operator ∇ is defined by the relation

$$\nabla f_x = f_x - f_{x-h}.$$

Prove (a) $\nabla = \dfrac{E-1}{E}$, (b) $\nabla E = \Delta = E\nabla$.

12. Form a table of "backward differences" ∇f_x of the function

$$x^3 - 3x^2 + 5x - 7$$

for $x = -1, 0, 1, 2, 3, 4, 5$. (Note that this table is identical with the table of "forward differences" Δf_x, but that the entries in the table have different names; $\nabla f_5 = \Delta f_4$, $\nabla f_4 = \Delta f_3$, and, in general, $\nabla f_x = \Delta f_{x-h}$.)

13. Prove the backward difference formulae

(a) $\nabla^n u_x = u_x - \binom{n}{1}u_{x-1} + \binom{n}{2}u_{x-2} - \binom{n}{3}u_{x-3} + \ldots ,$

(b) $u_x = 1 + \binom{x}{1}\nabla u_0 + \binom{x+1}{2}\nabla^2 u_0 + \binom{x+2}{3}\nabla^3 u_0 + \ldots$

14. From the following table of reciprocals, compute the reciprocal of 116.

x	100	110	120	130
$10^3 x^{-1}$	10.00000	9.09091	8.33333	7.69231

Why does the computed value differ from the tabular value 8.62069? Recompute the value for 116 when given that the entry for 140 is 7.14286.

15. An operator θ is called a *linear operator* if it satisfies the two conditions
$$\theta(f_x + g_x) = \theta f_x + \theta g_x, \quad \theta(c f_x) = c\, \theta f_x \quad (c \text{ a constant}).$$
Formulae (1.4) and (1.5) express the fact that Δ is a linear operator. Prove that the operator E and the operator ∇ (cf. Exercise 11) are linear operators.

16. The differential operator D is defined by the law $Df(x) = f'(x)$. Prove that D is a linear operator.

17. In virtue of (1.13), the operator $E^{1/2}$ is defined by the relation $E^{1/2}f(x) = f(x + \frac{1}{2}h)$. Determine whether or not $E^{1/2}$ is a linear operator.

18. Prove the following operator relations:
   (a) $(1 + \Delta)(1 - \nabla) = 1$, (b) $\nabla = 1 - E^{-1}$,
   (c) $\Delta\nabla = \Delta - \nabla$.

3. THE PROBLEM OF INTERPOLATION

Suppose that a function is defined by a table of values or, as is frequently the case in Applied Mathematics, is tabulated for a number of equidistant values of the argument (cf. the logarithmic function); then we can form a difference table and use the advancing difference formula (1.12) to interpolate further values. Whether or not these interpolated values are accurate ones will depend on the particular case. Before attempting to state any general principles, let us examine some examples.

EXAMPLE 1.2. Values $u_0 = -5$, $u_1 = -10$, $u_2 = -9$, $u_3 = 4$, $u_4 = 35$, of the function $u_x = x^3 - 6x - 5$ are given.

x	u_x	Δu_x	$\Delta^2 u_x$	$\Delta^3 u_x$
0	-5			
		-5		
1	-10		6	
		1		6
2	-9		12	
		13		6
3	4		18	
		31		
4	35			

In this example, $\Delta^n u_x = 0$ $(n > 3)$, and any value of u_x, within or without the confines of the table, can be found. In other words, interpolation (or extrapolation) is exact, because we know that we are dealing with a polynomial function.

EXAMPLE 1.3. Values of $u_x = \tan x$ are given for $x = 35°(2)45°$. The table is then used for interpolation at 40° and for extrapolation at 47°, 49°, 51°, etc. (For convenience, we adopt the usual convention of omitting the decimal point.)

$x°$	u_x	Δu_x	$\Delta^2 u_x$	$\Delta^3 u_x$	$\Delta^4 u_x$	$\Delta^5 u_x$
35	70021					
		5334				
37	75355		289			
		5623		39		
39	80978		328		5	
		5951		44		4
41	86929		372		9	
		6323		53		4
43	93252		425		13	
		6748		66		4
45	100000		491		17	
		7239		83		4
47	107239		574		21	
		7813		104		
49	115052		678			
		8491				
51	123543					

$$u_{40} = E^{1/2}u_{39} = \left[1 + \tfrac{1}{2}\Delta - \tfrac{1}{8}\Delta^2 + \tfrac{1}{16}\Delta^3 - \tfrac{5}{128}\Delta^4 + \tfrac{7}{256}\Delta^5\right] u_{39}$$
$$= .80978 + .029755 - .000465 + .000033 - .000005 + .000001$$
$$= .83910.$$

The tabular values for $\tan x$ are $u_{40} = .83910$, $u_{47} = 1.07237$, $u_{49} = 1.15037$, $u_{51} = 1.23490$.

In this example, we have assumed that fifth differences are constant; this is equivalent to replacing the given function $\tan x$ by a quintic polynomial. Such an approximating polynomial serves as an excellent representation of the function *within* the limits of the table; our given data tie down the approximating polynomial so closely that it agrees extremely well with the function $u_x = \tan x$ *within* the table limits. In other words, if $g(x)$ is the quintic polynomial having $g(x) = \tan x$ at the values $35°(2)45°$, then $g(x) - \tan x$ is nearly zero for any x in the range $35°-45°$. But, as soon as we pass *beyond* the limits of the table, when we extrapolate rather than interpolate, the divergence between $\tan x$ and its quintic polynomial approximation $g(x)$ becomes marked (and the divergence increases as we pass farther and farther from the tabulated values).

EXAMPLE 1.4. $u_x = e^x$

x	u_x	Δu_x	$\Delta^2 u_x$	$\Delta^3 u_x$	$\Delta^4 u_x$
0	1.000				
		1.718			
1	2.718		2.953		
		4.671		5.073	
2	7.389		8.026		8.716
		12.697		13.789	
3	20.086		21.815		
		34.512			
4	54.598				

In Example 1.3, the higher differences became relatively small, and exerted almost no influence on the interpolated functional values. In this example, it is clear not only that higher differences can not be considered constant, but that they are increasing rapidly. When we attempt to interpolate for $u_{1/2}$, we obtain

$$u_{1/2} = E^{1/2}u_0 = [1 + \tfrac{1}{2}\Delta - \tfrac{1}{8}\Delta^2 + \tfrac{1}{16}\Delta^3 - \tfrac{5}{128}\Delta^4]u_0$$

$$= 1.000 + .859 - .3691 + .3171 - .3405$$

$$= 1.467.$$

The correct value is 1.649. So we see that in this example not even interpolation is valid; indeed, Newton's Advancing Difference Formula is producing a non-convergent series. The given functional values are so coarsely spaced that the approximating polynomial does not begin to resemble the given function.

In conclusion, we draw together the following observations concerning interpolation.
1) If it is known that the function represented is a polynomial, that is, if the differences ultimately vanish, then both interpolation and extrapolation are exact (cf. Example 1.2).
2) If the function has differences which decrease quite rapidly and become approximately zero, then it can be represented to a very close approximation (within the range of the table) by a polynomial; interpolation is accurate, but extrapolation is untrustworthy if we go very far past the table limits.
3) If the function has a diverging sequence of differences, then even interpolation may fail (unless the differences increase so slowly that their effect is overbalanced by the binomial coefficients in the interpolation formula).
Further suggestive examples are given in the exercises.

EXERCISES

1. Let $u_x = \sin x$ be given for $x = 35°, 40°, 45°$; $u_{35} = .57358$, $u_{40} = .64279$, $u_{45} = .70711$. Write down the second-degree polynomial approximation to $\sin x$ in this region, and use it to give $u_{37.5}$, u_{90}, u_{180}. Compare with the tabular values $u_{37.5} = .60876$, $u_{90} = 1$, $u_{180} = 0$. Illustrate your results by making a large-scale graph of $\sin x$ and the approximating polynomial on the same set of axes.

2. Tabulate $u_x = 5^x$ from $x = 0(1)4$, and show that interpolation fails for $x = \frac{1}{2}$.

3. The following values of $\sin x°$ are known from elementary trigonometry.

x	$\sin x$
0	.00000
15	.25882
30	.50000
45	.70711
60	.86603
75	.96593
90	1.00000

Using differences of u_{30}, show that the above meagre information is sufficient to compute $\sin 37°30' = .60876$.

4. From the following table, compute $e^{-\pi/2} = e^{-1.5708}$. [This is the principal value of $i^i = (e^{i\pi/2})^i$.]

x	e^x
−1.55	.212248
−1.56	.210136
−1.57	.208045
−1.58	.205975
−1.59	.203926

Explain why interpolation succeeds here, but failed in Example 1.4 with the same function e^x.

5. From the following table of $u_x = \log_{10} x$, compute $\log_{10} 1.0025$ (the tabulated value is .0010844). Work from 1.002.

x	$\log_{10} x$
1.000	.0000000
1.001	.0004341
1.002	.0008677
1.003	.0013009
1.004	.0017337
1.005	.0021661

Now, again working from 1.002, compute the extrapolated value for log 1.022 (tabulated value .0094509). Do the same for 1.042 (tabulated value .0174507). Draw a large-scale graph from $x = 1.000$ to $x = 1.042$, and mark on it the function $\log_{10} x$ and the approximating polynomial.

6. Use the following table of $f_x = \tan x$

x	85°	86°	87°	88°	89°
f_x	11.430	14.301	19.081	28.636	57.290

to interpolate for $\tan 85°30'$ and for $\tan 87°30'$. Note how poorly the values compare with the tabular values 12.706 and 22.904.

In cases like this, where f_x changes very rapidly, the reciprocal function $g_x = (f_x)^{-1}$ will change more slowly, and *reciprocal interpolation* may be useful. Form a difference table for g_x in this problem (second differences of g_x are effectively constant), and use it to obtain $g(85°30')$ and $g(87°30')$; thence obtain excellent values of f by reciprocating.

7. Show that the method of Exercise 6 fails in Example 1.4 (the result of reciprocal interpolation is not quite as inaccurate as the result of direct interpolation, but the differences of the function $(u_x)^{-1}$ still do not decrease very rapidly).

4. THE PROBLEM OF SUBTABULATION

At times one is faced with the necessity of finding a large number of interpolated values for equally-spaced arguments. Thus the table might give u_x for $x = 0(5)30$, and it might be desired to compute u_x for $x = 0(1)30$. In such a case, interpolation by the formula

$$u_x = (1 + \Delta)^x u_0$$

becomes cumbersome. In practice, one usually needs to subtabulate only for the cases when the finer subdivision is $\frac{1}{2}$, $\frac{1}{5}$, or $\frac{1}{10}$ the length of the original division. The method, however, is quite general.

EXAMPLE 1.5.

$x°$	$u_x = \cos x$	Δu_x	$\Delta^2 u_x$	$\Delta^3 u_x$	$\Delta^4 u_x$
30	.86603				
		−4688			
35	.81915		−623		
		−5311		41	
40	.76604		−582		2
		−5893		43	
45	.70711		−539		
		−6432			
50	.64279				

We are required to tabulate $\cos x$ for $x = 30(1)50$. Obviously, the work involved by interpolating for each value required would be considerable; so let us introduce a new operator θ which has the property

$$\theta u_x = u_{x+h/5} - u_x.$$

Thus θ moves the functional value on by $1°$ only, whereas Δ moves the function by jumps of 5 degrees.

Consequently,

$$(1 + \theta)u_x = u_{x+h/5}; \qquad (1 + \theta)^5 u_x = u_{x+h}.$$

Thus we see that $(1 + \theta)^5 = 1 + \Delta$. Solving,

$$\theta = (1 + \Delta)^{1/5} - 1 = \frac{\Delta}{5} - \frac{2}{25}\Delta^2 + \frac{6}{125}\Delta^3 - \frac{21}{625}\Delta^4,$$

$$\theta^2 = \frac{\Delta^2}{25} - \frac{4}{125}\Delta^3 + \frac{16}{625}\Delta^4,$$

$$\theta^3 = \frac{\Delta^3}{125} - \frac{6}{625}\Delta^4,$$

$$\theta^4 = \frac{\Delta^4}{625}.$$

In general, of course, these expressions giving powers of θ in terms of Δ are infinite series; in this particular problem, they terminate with Δ^4 since we assume $\Delta^5 = 0$, that is, we represent u_x by a quartic polynomial.

Substituting, we obtain

$$\theta u_{30} = -.009376 + .0004984 + .000019\,224 - .000000\,672$$
$$= -.008859048$$

$$\theta^2 u_{30} = -.000261504, \quad \theta^3 u_{30} = .000003012, \quad \theta^4 u_{30} = .000000032.$$

We can then build up the complete table for u_x, using θ as our difference operator. Note that we keep more than five decimals in the higher differences, since round-off errors accumulate. In this problem it is convenient to keep nine decimals, since then there is no round-off at all. However, in general, we would keep six decimals in u_x, seven in θu_x, eight in $\theta^2 u_x$, etc.

x	u_x	θu_x	$\theta^2 u_x$	$\theta^3 u_x$	$\theta^4 u_x$
30	866030000				
		−8859048			
31	857170952		−261504		
		−9120552		3012	
32	848050400		−258492		32
		−9379044		3044	
33	838671356		−255448		32
		−9634492		3076	
34	829036864		−252372		
		−9886864			
35	819150000				

At the end, *and only then*, we round the answers to .85717, .84805, .83867, .82904. It is instructive to note the result of incorrectly rounding to five decimals in intermediate calculations.

x	u_x	θu_x	$\theta^2 u_x$	$\theta^3 u_x$
30	86603			
		−886		
31	85717		−26	
		−912		0
32	84805		−26	
		−938		0
33	83867		−26	
		−964		0
34	82903		−26	
		−990		
35	81913			

In this case, we have already lost two units in the fifth place by rounding off (and most examples will produce far more serious discrepancies). Clearly the round-off error would become quite large by the time the end of the table was reached. Furthermore, reproduction of the nominal zeros in u_{35} does provide a check on our work.

In concluding this section, we should note that ordinary advancing differences are largely superseded by central differences when it comes to practical interpolation, and that the subtabulation method of the present section is less desirable than the method using Everett's central difference formula, which will be discussed in Chapter III. (However, there are cases when we must use the method of this section if, as happens near the boundaries of a table, central differences are not available.) The present section is introduced here in order to indicate the nature of the problem, to provide further experience in working with operators, and to point up the desirability of maintaining extra nominal figures in intermediate computations in order to guard against round-off errors and to provide a check.

EXERCISES

1. Let u_x be a function for which Δ^3 may be considered constant. Develop formulae for subtabulation by means of the difference operator θ in the cases

(a) $\theta u_x = u_{x+h/2} - u_x$,   (b) $\theta u_x = u_{x+h/10} - u_x$.

2. Tabulate u_x for $x = 30(1)40$.

x	$\sin x$
30	.50000
32	.52992
34	.55919
36	.58778
38	.61566
40	.64279

3. The function

$$K(\alpha) = \int_0^{\pi/2} \frac{d\phi}{\sqrt{1 - \sin^2 \alpha \sin^2 \phi}}$$

is tabulated below.

$\alpha°$	$K(\alpha)$
0	1.5708
5	1.5738
10	1.5828
15	1.5981
20	1.6200

Tabulate $K(\alpha)$ for $\alpha = 0(1)5$.

4. From the following table of sines at $10°$ intervals, compute a table of sines for $x = 0(1)10$.

$x°$	$\sin x$
0	.00000
10	.17365
20	.34202
30	.50000
40	.64279

5. Compute e^x for $x = 0(.01).1$, given the values

x	e^x
0	1.0000
.1	1.1052
.2	1.2214
.3	1.3499
.4	1.4918

6. In Example 1.5, the expressions for θ, θ^2, ... , in terms of Δ, Δ^2, ... , were developed on the assumption that Δ^4 was constant. Obtain formulae for θ, θ^2, ... , in case (a) Δ^3 is constant, (b) Δ^5 is constant.

CHAPTER TWO

Divided Differences

1. DEFINITION AND NOTATION FOR DIVIDED DIFFERENCES

Suppose that the function u_x is given for values $x = a, b, c, d, \ldots$, where the intervals $b - a, c - b, d - c, \ldots$ are not necessarily equal (note that we do not demand that $a, b, c, d, \ldots$ be arranged in ascending order of magnitude, although this will usually be the case). Then we define the divided difference of u_a at b by the equation

$$(2.1) \qquad \underset{b}{\Delta} u_a = \frac{u_b - u_a}{b - a}.$$

The difference table may then be formed as before.

x	u_x	Δu_x	$\Delta^2 u_x$	$\Delta^3 u_x$
a	u_a			
		$\underset{b}{\Delta u_a}$		
b	u_b		$\underset{bc}{\Delta^2 u_a}$	
		$\underset{c}{\Delta u_b}$		$\underset{bcd}{\Delta^3 u_a}$
c	u_c		$\underset{cd}{\Delta^2 u_b}$	
		$\underset{d}{\Delta u_c}$		
d	u_d			

25

It might be noted that the notation $\Delta u_a \atop b$ for divided differences used in (2.1) is not universal. Several other notations are in use, and all have their proponents (and their disadvantages). The student will find a list of alternative notations in W. E. Milne, *Numerical Calculus*, Appendix A.

It is instructive to compute the higher divided differences in terms of functional values. Thus

$$\Delta^2_{bc} u_a = \frac{\Delta_c u_b - \Delta_b u_a}{c - a}$$

$$= \frac{\dfrac{u_c - u_b}{c - b} - \dfrac{u_b - u_a}{b - a}}{c - a}$$

$$= \frac{u_a}{(a - b)(a - c)} + \frac{u_b}{(b - c)(b - a)} + \frac{u_c}{(c - a)(c - b)}.$$

Similarly,

$$\Delta^3_{bcd} u_a = \frac{\Delta^2_{cd} u_b - \Delta^2_{bc} u_a}{d - a}$$

$$= \frac{u_a}{(a - b)(a - c)(a - d)} + \ldots + \frac{u_d}{(d - a)(d - b)(d - c)}.$$

These two results are special cases of the general

THEOREM 2.1. If $a, b, c, \ldots$ are values of the argument x, then

$$\Delta^r_{bc \ldots jk} u_a = \frac{u_a}{(a - b)(a - c) \ldots (a - k)} + \ldots + \frac{u_k}{(k - a)(k - b) \ldots (k - j)}.$$

The proof consists of an obvious induction, and so is omitted.

EXAMPLE 2.1. Build up a divided difference table, given that

$$u_{-2} = 5, \quad u_0 = 3, \quad u_3 = 15, \quad u_4 = 47, \quad u_9 = 687.$$

x	u_x	Δu_x	$\Delta^2 u_x$	$\Delta^3 u_x$
-2	5			
		-1		
0	3		$+1$	
		$+4$		$+1$
3	15		$+7$	
		$+32$		$+1$
4	47		$+16$	
		$+128$		
9	687			

It might be noted that the triangular arrangement of a divided difference table is extremely convenient; we have

$$\underset{3,4}{\Delta^2 u_0} = \frac{32 - 4}{4 - 0},$$

and the triangle drawn immediately points to the proper divisor $4 - 0$.

EXAMPLE 2.2. Build up a divided difference table, given that

$$u_3 = 15, \quad u_{-2} = 5, \quad u_9 = 687, \quad u_0 = 3, \quad u_4 = 47.$$

(Obviously, we are dealing with the same function as in Example 2.1.)

x	u_x	Δu_x	$\Delta^2 u_x$	$\Delta^3 u_x$
3	15			
		+2		
-2	5		+1	
		-1		+1
0	3		+7	
		+76		+1
9	687		+13	
		+128		
4	47			

We note that in both cases the third divided differences of the function (actually $x^3 - 5x + 3$) are constant; this is a special case of

THEOREM 2.2. If u_x is a polynomial of degree n, then $\Delta^n u_x$ is constant.

Proof.

$$\underset{x+h}{\Delta} x^n = \frac{(x + h)^n - x^n}{x + h - x} = \frac{nhx^{n-1} + \dots}{h}$$

$$= \text{a polynomial of degree } n - 1.$$

Now Δ is a linear operator, that is, $\Delta(f + g) = \Delta f + \Delta g, \ \Delta cf = c\Delta f$ (c constant). Hence it follows that the first divided difference of

$$u_x = a_0 + a_1 x + \dots + a_n x^n$$

is a polynomial of degree $n - 1$. Consequently, the second divided difference is a polynomial of degree $n - 2$, the nth divided difference is constant, and all higher divided differences are zero.

We conclude this section by stating an important corollary of Theorem 2.1 as

THEOREM 2.3. (*The Symmetry Property*). If a divided difference $\underset{bc \dots jk}{\Delta^r u_a}$ is given, then it is unaltered by any permutation of the letters $a, b, c, \dots, j, k$.

The proof of Theorem 2.3 is immediate, when we note that the expression in Theorem 2.1 is a symmetric function of all $r + 1$ letters $a, b, \ldots, j, k$. As an illustration, we note that in Example 2.1 we found $\Delta^2 u_{-2} = 1$; in Example 2.2, $\Delta^2 u_3 = 1$.
$\underset{-2,0}{\Delta^2 u_3}$ $\underset{0,3}{}$

EXERCISES

1. Construct a divided difference table, using the data of Example 2.1, with arguments in the order 3, -2, 9, 0, 4.

2. From the values $u_{-1} = 1$, $u_1 = -3$, $u_4 = 21$, $u_6 = 127$, form a table giving $\underset{4,6}{\Delta^2 u_1}$. Form another table using the order u_4, u_1, u_6, u_{-1}, and read off the value of $\underset{1,6}{\Delta^2 u_4}$.

3. Suppose divided differences are taken at arguments $a, b, c, \ldots, j, k$ with $k - j = \ldots = c - b = b - a = h$ (h constant). Prove by induction that

$$\underset{bc\ldots k}{\Delta^r u_a} = \frac{\Delta^r u_a}{r!\, h^r}.$$

Deduce that $\Delta^n(a_0 + a_1 x + \ldots + a_n x^n) = a_n$.

4. Make an ordinary and a divided difference table for the data

x	0	2	4	6
u_x	-18	6	55	104

5. Construct a divided difference table given

$x°$	$u_x = \sin x$
30	.50000
33	.54464
34	.55919
40	.64279

(Higher differences should be carried to additional decimals.)

2. NEWTON'S DIVIDED DIFFERENCE FORMULA

Consider the function u_x for arguments $x, a, b, c, d, \ldots, j, k$; then

$$\underset{a}{\Delta u_x} = \underset{x}{\Delta u_a} = \frac{u_x - u_a}{x - a},$$

and, solving for u_x, we obtain

(2.2) $u_x = u_a + (x - a)\underset{x}{\Delta u_a}.$

From the symmetry property, we obtain

$$\underset{bx}{\Delta^2} u_a = \underset{ab}{\Delta^2} u_x = \frac{\underset{a}{\Delta u_x} - \underset{b}{\Delta u_a}}{x - b};$$

thence, using the symmetry property,

$$\underset{x}{\Delta u_a} = \underset{b}{\Delta u_a} + (x - b)\underset{bx}{\Delta^2} u_a,$$

and, substituting in (2.2), we obtain

(2.3) $$u_x = u_a + (x - a)\underset{b}{\Delta u_a} + (x - a)(x - b)\underset{bx}{\Delta^2} u_a.$$

Finally,

$$\underset{bcx}{\Delta^3} u_a = \underset{abc}{\Delta^3} u_x = \frac{\underset{ab}{\Delta^2} u_x - \underset{bc}{\Delta^2} u_a}{x - c}$$

and

$$\underset{bx}{\Delta^2} u_a = \underset{bc}{\Delta^2} u_a + (x - c)\underset{bcx}{\Delta^3} u_a;$$

substitution in (2.3) gives the expression

(2.4) $$u_x = u_a + (x - a)\underset{b}{\Delta u_a} + (x - a)(x - b)\underset{bc}{\Delta^2} u_a$$

$$+ (x - a)(x - b)(x - c)\underset{bcx}{\Delta^3} u_a.$$

Continue this process until nth differences are reached; under the assumption that u_x is represented by an nth degree polynomial, all higher differences vanish and we have *Newton's Divided Difference Formula*

(2.5) $$u_x = u_a + A\,\underset{b}{\Delta u_a} + AB\,\underset{bc}{\Delta^2} u_a + ABC\,\underset{bcd}{\Delta^3} u_a + \ldots + ABC \ldots J\underset{bc \ldots k}{\Delta^n} u_a,$$

where there are $n + 1$ arguments $a, b, c, \ldots, k$, and where we use the abbreviations $x - a = A$, $x - b = B$, $\ldots$, $x - k = K$.

If the arguments $a, b, c, d, \ldots$ are taken as $0, 1, 2, \ldots$, then $\Delta^r u_a = \Delta^r u_0/r!$, and (2.5) specializes to the result

$$u_x = u_0 + \binom{x}{1}\Delta u_0 + \binom{x}{2}\Delta^2 u_0 + \ldots + \binom{x}{n}\Delta^n u_0,$$

which is just the ordinary advancing difference formula

$$u_x = E^x u_0 = (1 + \Delta)^x u_0$$

of Chapter I.

Newton's formula (2.5), as well as all other divided difference formulae, is most easily remembered by a mnemonic known as "Sheppard's Zigzag Rule". This will now be explained.

Suppose we are given a table of divided differences (Table 2.1 illustrates the case of six arguments a, b, c, d, e, f). Draw *any* zigzag line connecting any argument u with the highest difference in the table. We may only zigzag from one difference to an adjacent difference of the next higher order (that is, at any stage we have a choice of two paths). The letters a, b, c, d, e, f will appear in the differences along the (solid) zigzag line in some order; the one illustrated is c, d, b, e, f, a.

Table 2.1.   SHEPPARD'S ZIGZAG RULE

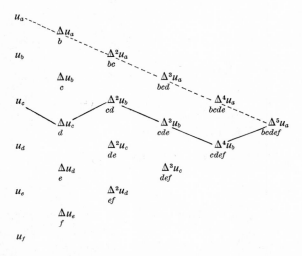

Sheppard's Rule is simply this: Expand u_x in terms of differences on the zigzag line using coefficients 1, C, D, B, E, F, that is, *the large letters appear in the same order as the small ones.* Thus, the preceding diagram represents the expansion

$$u_x = u_c + C \underset{d}{\Delta u_c} + C D \underset{cd}{\Delta^2 u_b} + C D B \underset{cde}{\Delta^3 u_b} + C D B E \underset{cdef}{\Delta^4 u_b} + C D B E F \underset{bcdef}{\Delta^5 u_a}.$$

The Newton Divided Difference Formula (2.5) corresponds to the dotted line in the figure, that is, to the successive divided differences of u_a and the coefficient order A, B, C, D, E. Clearly, a great many formulae can be read off from Table 2.1; all of them can be rigorously established by the method used in obtaining Formula (2.5).

EXERCISES

1. Use Newton's formula to obtain a polynomial approximation to the data

$$u_{10} = 355, \quad u_0 = -5, \quad u_8 = -21, \quad u_1 = -14, \quad u_4 = -125.$$

2. Interpolate for tan 31° given

x	$\tan x$
30°	.57735
30°30′	.58905
32°	.62487
32°12′	.62973

3. Find u_{28} given

$$u_{20} = 24.37, \quad u_{22} = 49.28, \quad u_{29} = 162.86, \quad u_{32} = 240.50.$$

3. THE LAGRANGE INTERPOLATION FORMULA

Lagrange's Formula is merely a variant of Newton's Formula (2.5); again let u_x be given for $n + 1$ values $a, b, c, \ldots, j, k$, and assume that u_x is represented by an nth degree polynomial. Then the $(n + 1)$th differences vanish; in particular,

$$\underset{abc \cdots k}{\Delta^{n+1}} u_x = 0.$$

Using Theorem 2.1, we can write this equation as

$$(2.6) \quad \frac{u_x}{(x - a)(x - b) \ldots (x - k)} + \frac{u_a}{(a - b)(a - c) \ldots (a - k)(a - x)}$$
$$+ \ldots + \frac{u_k}{(k - a)(k - b) \ldots (k - j)(k - x)} = 0.$$

Multiply through by $(x - a)(x - b) \ldots (x - k)$, and we obtain

$$(2.7)$$
$$u_x = \frac{(x - b)(x - c) \ldots (x - k)}{(a - b)(a - c) \ldots (a - k)} u_a + \ldots + \frac{(x - a)(x - b) \ldots (x - j)}{(k - a)(k - b) \ldots (k - j)} u_k.$$

This expression (2.7) is called Lagrange's Interpolation Formula; it expresses u_x in terms of the $n + 1$ given functional values $u_a, u_b, \ldots, u_k$.

Both Formula (2.5) and Formula (2.7) are of more theoretical than practical importance. Thus (2.7) is intimately connected with the partial fraction break-up of a rational function. For instance, let $F(x) = u(x)/v(x)$ be a given rational function, where $u(x)$ has degree not greater than n, where $v(x) = (x - a)(x - b) \ldots (x - k)$, and the zeros of $v(x)$ are distinct. From (2.6),

$$F(x) = \frac{u(x)}{(x - a)(x - b) \ldots (x - k)}$$
$$= \frac{u(a)}{(a - b)(a - c) \ldots (a - .k)} \frac{1}{x - a} + \ldots$$
$$+ \frac{u(k)}{(k - a)(k - b) \ldots (k - j)} \frac{1}{x - k}.$$

This result can be written in the form

(2.8) $$F(x) = \frac{u(a)}{v'(a)}\frac{1}{x-a} + \dots + \frac{u(k)}{v'(k)}\frac{1}{x-k},$$

which is the usual form of the partial fraction decomposition of $F(x)$.

EXERCISES

1. Use the Lagrange Interpolation Formula to compute u_4, given

$$u_0 = 707, \quad u_2 = 819, \quad u_3 = 866, \quad u_6 = 966.$$

Repeat, using divided differences.

2. Use Lagrange's Formula to express as sums of partial fractions:

(a) $\dfrac{x^2 + 6x + 1}{(x^2 - 1)(x - 4)(x - 6)}$ (b) $\dfrac{3x^2 + x + 1}{x^3 - 6x^2 + 11x - 6}$

(c) $\dfrac{x^2 + 2x + 4}{x^2 - x + 12}$ (d) $\dfrac{x^3 - 4x^2 + 21x - 8}{x^3 - 6x^2 + 11x - 6}.$

3. Suppose that $F(x) = u(x)/v(x)$, where

$$v(x) = (x - a)^2(x - c)(x - d) \dots (x - k).$$

Obtain the analogue of Formula (2.8).

CHAPTER THREE

Central Differences

1. NOTATION FOR CENTRAL DIFFERENCES

Consider a function u_x tabulated for equally-spaced arguments a, $a + h$, $a + 2h$, ... It frequently happens that we are interested in the behaviour of u_x in the immediate neighbourhood of $a + rh$; with this in view, it becomes natural to make the transformation $X = \dfrac{x - (a + rh)}{h}$. The value u_{a+rh} now becomes u_0, and the interval of differencing becomes 1. The difference table takes on the following appearance (we write u_x for u_X, etc.).

x	X	u_x	Δu_x	$\Delta^2 u_x$	$\Delta^3 u_x$	$\Delta^4 u_x$
$a + (r - 2)h$	-2	u_{-2}				
			Δu_{-2}			
$a + (r - 1)h$	-1	u_{-1}		$\Delta^2 u_{-2}$		
			Δu_{-1}		$\Delta^3 u_{-2}$	
$a + rh$	0	u_0		$\Delta^2 u_{-1}$		$\Delta^4 u_{-2}$
			Δu_0		$\Delta^3 u_{-1}$	
$a + (r + 1)h$	1	u_1		$\Delta^2 u_0$		
			Δu_1			
$a + (r + 2)h$	2	u_2				

The numbers appearing in this difference table are, of course, independent of the nomenclature; so we might justifiably refer to them as *Central Differences*.

However, we normally introduce a special *central difference operator* δ by the definition

(3.1) $$\delta = E^{1/2} - E^{-1/2}.$$

It immediately follows that

(3.2) $$\delta E^{1/2} = E - 1 = \Delta.$$

Hence, in δ notation, we can write

$$\Delta u_{-2} = \delta E^{1/2} u_{-2} = \delta u_{-3/2}; \qquad \Delta u_{-1} = \delta u_{-1/2}; \qquad \Delta u_0 = \delta u_{1/2}.$$

Second differences follow from

(3.3) $$\delta^2 E = \Delta^2.$$

Thus $$\Delta^2 u_{-2} = \delta^2 E u_{-2} = \delta^2 u_{-1}; \qquad \Delta^2 u_{-1} = \delta^2 u_0; \qquad \text{etc.}$$

Similarly,

(3.4) $$\delta^3 E^{3/2} = \Delta^3,$$

and $$\Delta^3 u_{-2} = \delta^3 u_{-1/2}, \qquad \Delta^3 u_{-1} = \delta^3 u_{1/2}.$$

It follows that in δ notation the difference table may be rewritten in the form of Table 3.1.

Table 3.1. CENTRAL DIFFERENCE NOTATION

X	u_x	δu_x	$\delta^2 u_x$	$\delta^3 u_x$	$\delta^4 u_x$
-2	u_{-2}				
		$\delta u_{-3/2}$			
-1	u_{-1}		$\delta^2 u_{-1}$		
		$\delta u_{-1/2}$		$\delta^3 u_{-1/2}$	
0	u_0		$\delta^2 u_0$		$\delta^4 u_0$
		$\delta u_{1/2}$		$\delta^3 u_{1/2}$	
1	u_1		$\delta^2 u_1$		
		$\delta u_{3/2}$			
2	u_2				

The great advantage of the δ notation is that, as manifest from the above table, $\delta^r u_a$ appears on the line exactly opposite u_a.

We conclude this section by introducing another operator

(3.5) $$\mu = \tfrac{1}{2}(E^{1/2} + E^{-1/2}).$$

We see that μ is an averaging operator; thus

$$\mu u_0 = \tfrac{1}{2}(u_{1/2} + u_{-1/2}).$$

Various relations exist among the operators $\Delta, E, \delta,$ and μ. Two of the most important are

(3.6) $$\mu^2 = \tfrac{1}{4}(E + E^{-1} + 2) = 1 + \frac{\delta^2}{4},$$

and

(3.7) $$\mu\delta = \tfrac{1}{2}(E - E^{-1}) = \tfrac{1}{2}\Delta E^{-1} + \tfrac{1}{2}\Delta.$$

Thus $$\mu\delta u_0 = \tfrac{1}{2}(\delta u_{1/2} + \delta u_{-1/2}) = \tfrac{1}{2}(\Delta u_{-1} + \Delta u_0).$$

EXERCISES

1. Two operators θ and ϕ commute if $\theta\phi = \phi\theta$. Check that μ, δ, E, Δ, and ∇ commute with one another.

2. Form a difference table, given

$$u_{-2} = 15, \quad u_{-1} = 12, \quad u_0 = 5, \quad u_1 = 0, \quad u_2 = 3.$$

Identify $\delta u_{1/2}$, $\delta^2 u_0$, $\delta^3 u_{-1/2}$ in the table, and give the other names for these differences in terms of Δ and ∇.

3. Prove that $\delta x = h$ (the interval of differencing).

4. Prove the following identities.

    (a) $\sqrt{1 + \delta^2 \mu^2} = 1 + \dfrac{\delta^2}{2}$,   (b) $E^{1/2} = \mu + \dfrac{\delta}{2}$,

    (c) $\nabla = \delta E^{-1/2}$,   (d) $E^{-1/2} = \mu - \dfrac{\delta}{2}$.

5. The derivative operator D is defined by $Du(x) = \dfrac{du(x)}{dx}$. Prove that D commutes with δ, E, Δ, μ, ∇.

2. THE GAUSS AND STIRLING FORMULAE

In this section, we shall use divided differences to develop two formulae due to Gauss. Actually, neither the divided difference formula nor the Gauss formulae are of great practical use; however, they serve as intermediate steps in obtaining the important formulae of the next sections, and provide us with an opportunity to increase our mastery of the δ notation.

Starting with a central difference table in Δ notation, we have

The table can, of course, be extended in either direction. Since the interval of differencing is constant and equal to unity, the divided differences of u_x are given by the relation

$$\Delta^r u_x = \frac{\Delta^r u_x}{r!} .$$

So, if we apply Sheppard's zigzag rule along the solid line, we obtain

$$u_x = u_0 + x\frac{\Delta u_0}{1!} + x(x-1)\frac{\Delta^2 u_{-1}}{2!} + x(x-1)(x+1)\frac{\Delta^3 u_{-1}}{3!}$$

$$+ x(x-1)(x+1)(x-2)\frac{\Delta^4 u_{-2}}{4!}$$

$$+ x(x-1)(x+1)(x-2)(x+2)\frac{\Delta^5 u_{-2}}{5!} + \ldots$$

Rearranging this expression gives the *Gauss Forward Formula*

$$(3.8) \quad u_x = u_0 + \binom{x}{1}\Delta u_0 + \binom{x}{2}\Delta^2 u_{-1} + \binom{x+1}{3}\Delta^3 u_{-1}$$

$$+ \binom{x+1}{4}\Delta^4 u_{-2} + \binom{x+2}{5}\Delta^5 u_{-2} + \ldots$$

On the other hand, if we use the Sheppard zigzag rule along the dotted line, and then rearrange the terms in a similar fashion, we have the *Gauss Backward Formula*

$$(3.9) \quad u_x = u_0 + \binom{x}{1}\Delta u_{-1} + \binom{x+1}{2}\Delta^2 u_{-1} + \binom{x+1}{3}\Delta^3 u_{-2}$$

$$+ \binom{x+2}{4}\Delta^4 u_{-2} + \binom{x+2}{5}\Delta^5 u_{-3} + \ldots$$

Changing these two formulae to δ notation produces respectively

$$(3.10) \quad u_x = u_0 + \binom{x}{1}\delta u_{1/2} + \binom{x}{2}\delta^2 u_0 + \binom{x+1}{3}\delta^3 u_{1/2}$$

$$+ \binom{x+1}{4}\delta^4 u_0 + \binom{x+2}{5}\delta^5 u_{1/2} + \ldots$$

and

$$(3.11) \quad u_x = u_0 + \binom{x}{1}\delta u_{-1/2} + \binom{x+1}{2}\delta^2 u_0 + \binom{x+1}{3}\delta^3 u_{-1/2}$$

$$+ \binom{x+2}{4}\delta^4 u_0 + \binom{x+2}{5}\delta^5 u_{-1/2} + \ldots$$

Thus we have expansions in terms of the central differences of u_0 and of either

$u_{1/2}$ or $u_{-1/2}$. If we add these two expressions (3.10) and (3.11), and divide by two, we immediately get *Stirling's Formula*

$$(3.12) \quad u_x = u_0 + \binom{x}{1}\mu\delta u_0 + \frac{x}{2}\binom{x}{1}\delta^2 u_0 + \binom{x+1}{3}\mu\delta^3 u_0$$

$$+ \frac{x}{4}\binom{x+1}{3}\delta^4 u_0 + \binom{x+2}{5}\mu\delta^5 u_0 + \dots$$

or, alternatively,

$$(3.13) \quad u_x = u_0 + \frac{x}{1!}\mu\delta u_0 + \frac{x^2}{2!}\delta^2 u_0 + \frac{x(x^2-1^2)}{3!}\mu\delta^3 u_0$$

$$+ \frac{x^2(x^2-1^2)}{4!}\delta^4 u_0 + \frac{x(x^2-1^2)(x^2-2^2)}{5!}\mu\delta^5 u_0 + \dots$$

It is convenient to note that the differences appearing in Stirling's Formula lie on a horizontal line through u_0, as illustrated in the following diagram

$$u_0 \text{——} \mu\delta u_0 \text{——} \delta^2 u_0 \text{——} \mu\delta^3 u_0 \text{——} \delta^4 u_0 \text{——} \dots$$

The even-ordered differences $\delta^{2n}u_0$ are actual tabular values; the odd-ordered mean differences

$$\mu\delta^{2n+1}u_0 = \tfrac{1}{2}(\delta^{2n+1}u_{-1/2} + \delta^{2n+1}u_{1/2})$$

are averages of adjacent tabular values.

We are now in a position to point out the superiority of central difference formulae. If we compute $u_{.2}$ by the advancing difference formula, we obtain (to fifth differences)

$$u_{.2} = E^{.2}u_0 = (1 + \Delta)^{.2}u_0$$

$$= u_0 + .2\Delta u_0 - .08\Delta^2 u_0 + .048\Delta^3 u_0 - .0336\Delta^4 u_0 + .025536\Delta^5 u_0.$$

Using Stirling's formula,

$$u_{.2} = u_0 + .2\mu\delta u_0 + .02\delta^2 u_0 - .032\mu\delta^3 u_0 - .0016\delta^4 u_0 + .006336\mu\delta^5 u_0.$$

It is clear that the coefficients in Stirling's formula decrease much more rapidly; this not only saves labour, but ensures that the contribution made by higher differences is much less. Since any uncertainties in the given data, arising either from experimental errors or from round-off errors, are enormously magnified in the higher differences, we prefer formulae in which the effects of these higher differences are small. There is the further point that the advancing difference formula uses only information supplied by u_0, u_1, u_2, u_3, u_4 (to fourth differences); Stirling's formula uses information supplied by u_0, u_1, u_{-1}, u_2, u_{-2} (to fourth differences). It seems logical to suppose that we get a better picture of the function's behaviour by employing values on both sides of u_0 rather than just those values following u_0.

EXAMPLE 3.1. Use Stirling's formula to compute $u_{12.2}$ from the following table ($u_x = 1 + \log_{10} \sin x$).

$x°$	X	$v_x = 10^5 u_x$	δv_x	$\delta^2 v_x$	$\delta^3 v_x$	$\delta^4 v_x$
10	-2	23967				
			4093			
11	-1	28060		-365		
			3728		58	
12	0	31788		-307		-13
			3421		45	
13	1	35209		-262		
			3159			
14	2	38368				

Taking 12 as origin, we get

$$v_{.2} = 31788 + .2\frac{7149}{2} + .02(-307) - .032\frac{103}{2} - .0016(-13)$$

$$= 31788 + 714.9 - 6.14 - 1.648 + 0.208$$

$$= 32495.$$

The value tabulated for $12°12'$ is .32495.

If we employ advancing differences, we get

$$v_{.2} = 31788 + .2(3421) - .08(-262) + .048(32) - .0336(-13)$$

$$= 31788 + 684.2 + 20.96 + 1.536 + .4368$$

$$= 32495.$$

In this case, we naturally obtain the same value. However, the higher differences contribute much more to the value and consequently could have had an exaggerated effect in the case of a less well-behaved function.

EXERCISES

1. Write out the Stirling and advancing difference formulae for $u_{.1}$. Use them in Example 3.1 to compute u_x for $x = 12°6'$.

2. From the following table of $u_x = 1.0025^x$, compute $1.0025^{77.15}$.

x	u_x
75	1.20594804
76	1.20896291
77	1.21198532
78	1.21501528
79	1.21805282

Use all three formulae of this section; in each case indicate on a line diagram the differences which are employed.

3. From the following table of $u_x = 10^3 x^{-1}$, compute $u_{431.75}$. (Note that $\delta^3 u_x$ is effectively zero, except for accumulating round-off.)

x	u_x
430	2.325581
431	2.320186
432	2.314815
433	2.309469
434	2.304147

4. The following table of functional values is given.

x	$10^5 \log x$
.010	−460517
.011	−450986
.012	−442285
.013	−434281
.014	−426870
.015	−419971
.016	−413517
.017	−407454
.018	−401738
.019	−396332

(a) Make a difference table as far as fifth differences.

(b) Why should we stop at fifth differences, even though they are not constant?

(c) Using the Stirling central difference formula, compute log .01525.

3. THE BESSEL FORMULA

Let us begin with the Gauss Forward Formula (3.10), and combine it with the Gauss Backward Formula based on a zigzag line just one unit below the line we previously used. The latter formula then expresses u_x in terms of the differences of $u_{1/2}$ and u_1, rather than in terms of the differences of $u_{-1/2}$ and u_0, and becomes

$$(3.14) \quad u_x = u_1 + \binom{x-1}{1} \delta u_{1/2} + \binom{x}{2} \delta^2 u_1 + \binom{x}{3} \delta^3 u_{1/2}$$

$$+ \binom{x+1}{4} \delta^4 u_1 + \binom{x+1}{5} \delta^5 u_{1/2} + \ldots;$$

averaging (3.14) with (3.10), we obtain *Bessel's Formula*

$$(3.15) \quad u_x = \mu u_{1/2} + (x - \tfrac{1}{2}) \delta u_{1/2} + \binom{x}{2} \mu \delta^2 u_{1/2} + \tfrac{1}{3}(x - \tfrac{1}{2})\binom{x}{2} \delta^3 u_{1/2}$$

$$+ \binom{x+1}{4} \mu \delta^4 u_{1/2} + \tfrac{1}{5}(x - \tfrac{1}{2})\binom{x+1}{4} \delta^5 u_{1/2} + \ldots$$

The differences and mean differences appearing in (3.15) all lie on the horizontal line through $u_{1/2}$.

Bessel's Formula, which is of very great importance, can be put in a more elegant form by setting $x = z + \frac{1}{2}$. We then obtain

$$(3.16) \quad u_{z+1/2} = \mu u_{1/2} + z\,\delta u_{1/2} + \frac{z^2 - \frac{1}{4}}{2!}\mu\delta^2 u_{1/2} + \frac{z(z^2 - \frac{1}{4})}{3!}\delta^3 u_{1/2}$$

$$+ \frac{(z^2 - \frac{1}{4})(z^2 - \frac{9}{4})}{4!}\mu\delta^4 u_{1/2} + \frac{z(z^2 - \frac{1}{4})(z^2 - \frac{9}{4})}{5!}\delta^5 u_{1/2} + \dots$$

Putting $z = 0$ gives the important formula, useful for halving the interval of tabulation,

$$(3.17) \qquad u_{1/2} = \mu u_{1/2} - \tfrac{1}{8}\mu\delta^2 u_{1/2} + \tfrac{3}{128}\mu\delta^4 u_{1/2} - \dots$$

We should note that, by proper choice of origin, x in (3.15) or any other central difference formula may be taken in the range $0 \leq x \leq 1$ or in the range $-\frac{1}{2} \leq x \leq +\frac{1}{2}$.

EXAMPLE 3.2. $u_x = x^{1/3}$; compute $344.5^{1/3}$, given the following values.

x	$v_x = 10^6 u_x$	δv_x	$\delta^2 v_x$
342	6993191		
		6809	
343	7000000		−13
		6796	
344	7006796		−13
		6783	
345	7013579		−13
		6770	
346	7020349		−13
		6757	
347	7027106		

Taking origin at 344, we immediately have

$$v_{1/2} = \frac{14020375}{2} - \frac{1}{8}(-13) = 7010187.5 + 1.6$$

$$= 7010189.$$

EXERCISES

1. Compute $\sqrt[3]{5863}$ given the following table of $u_x = x^{1/3}$.

x	u_x
5600	17.75808
5700	17.86316
5800	17.96702
5900	18.06969
6000	18.17121

2. Use Bessel's Formula to compute $\log_{10} 1162.5$ and $\log_{10} 1169$ from the table

x	$\log_{10} x$
1150	3.0606978
1155	3.0625820
1160	3.0644580
1165	3.0663259
1170	3.0681859
1175	3.0700379
1180	3.0718820

3. The function $\log E$, where $E = \int_0^{\pi/2} \sqrt{1 - \sin^2 \alpha \sin^2 \phi}\, d\phi$, is tabulated below.

$\alpha°$	$\log E$
0	.196120
5	.195293
10	.192815
15	.188690
20	.182928

Compute $\log E$ for $\alpha = 9°$ and $12°30'$.

4. The function $u_x = 1.015^{-x}$ is tabulated below.

x	u_x
48	.48936170
50	.47500468
52	.46106887
54	.44754192
56	.43441182

Compute 1.015^{-53} by Bessel's formula (tabulated value .45425505).

5. Suppose Formula (3.17) is written

$$u_{1/2} = \sum_{r=0}^{\infty} (-)^r a_r \mu \delta^{2r} u_{1/2}.$$

Prove that

$$\frac{a_{r+1}}{a_r} = \frac{2r+1}{8(r+1)}.$$

Use this relation to compute the first six a_i to eight decimals.

6. (Comrie's Throwback) Show that the second and fourth difference terms in Bessel's Formula can be written in the form

$$\binom{x}{2}[\mu\delta^2 u_{1/2} - C\mu\delta^4 u_{1/2}],$$

where C is a function of x such that $.1667 \leq C \leq .1875$ in the range $0 \leq x \leq 1$. Supposing that C is taken equal to a constant K, prove the following results.
(a) The error in so doing is given by

$$g(x) = \tfrac{1}{2}x(x-1)[K + \tfrac{1}{12}(x+1)(x-2)].$$

(b) $g(x)$ has three extrema in the range $0 \leq x \leq 1$.
(c) $g(x)$ has smallest maximal variation from zero for

$$K = \frac{3+\sqrt{2}}{24} = .1839255.$$

(d) For this choice of K, the value in (a) is .000448.
(e) If second differences $\delta^2 u_x$ are replaced by "modified second differences"

$$\delta_m^2 u_x = \delta^2 u_x - K\delta^4 u_x,$$

then the error introduced by using $\delta_m^2 u_x$ and neglecting the fourth difference term is less than $\tfrac{1}{2}$ unit in the last figure, provided that $\delta^4 u_x < 1100$.

4. THE EVERETT FORMULAE

We require

LEMMA 3.1.

$$\binom{z}{j}\delta^j u_a + \binom{z+1}{j+1}\delta^{j+1}u_{a+1/2} = \binom{z+1}{j+1}\delta^j u_{a+1} - \binom{z}{j+1}\delta^j u_a.$$

Proof. The left-hand side becomes

$$\left[\binom{z+1}{j+1} - \binom{z}{j+1}\right] \delta^j u_a + \binom{z+1}{j+1} \delta^{j+1} u_{a+1/2}$$

$$= \binom{z+1}{j+1} [\delta^j (1 + \delta E^{1/2}) u_a] - \binom{z}{j+1} \delta^j u_a$$

$$= \binom{z+1}{j+1} \delta^j u_{a+1} - \binom{z}{j+1} \delta^j u_a.$$

Now apply this identity to the Gauss Forward Formula, grouping together the first and second, third and fourth, fifth and sixth terms. We obtain

(3.18) $u_x = \binom{x}{1} u_1 + \binom{x+1}{3} \delta^2 u_1 + \binom{x+2}{5} \delta^4 u_1 + \cdots$

$$- \binom{x-1}{1} u_0 - \binom{x}{3} \delta^2 u_0 - \binom{x+1}{5} \delta^4 u_0 - \cdots$$

This is known as *Everett's First Formula*; it expands u_x in terms of the even differences lying on the two horizontal lines through u_0 and u_1. If we put $x + y = 1$, we obtain a more symmetric and convenient form

(3.19) $u_x = x\left[u_1 + \dfrac{x^2 - 1}{3!} \delta^2 u_1 + \dfrac{(x^2 - 1)(x^2 - 4)}{5!} \delta^4 u_1 + \cdots\right]$

$$+ y\left[u_0 + \dfrac{y^2 - 1}{3!} \delta^2 u_0 + \dfrac{(y^2 - 1)(y^2 - 4)}{5!} \delta^4 u_0 + \cdots\right].$$

If we take the Gauss Backward Formula and group the second and third, fourth and fifth, sixth and seventh terms, we get (again applying Lemma 3.1)

(3.20) $u_x = u_0 + \binom{x+1}{2} \delta u_{1/2} + \binom{x+2}{4} \delta^3 u_{1/2} + \binom{x+3}{6} \delta^5 u_{1/2} + \cdots$

$$- \binom{x}{2} \delta u_{-1/2} - \binom{x+1}{4} \delta^3 u_{-1/2} - \binom{x+2}{6} \delta^5 u_{-1/2} + \cdots$$

This is *Everett's Second Formula*; it expands u_x in terms of the odd differences lying on the two horizontal lines through $u_{1/2}$ and $u_{-1/2}$. Using $p = \frac{1}{2} + x$, $q = \frac{1}{2} - x$, we can put (3.20) in the form

(3.21) $u_x = u_0 + \left[\dfrac{p^2 - \frac{1}{4}}{2!} \delta u_{1/2} + \dfrac{(p^2 - \frac{1}{4})(p^2 - \frac{9}{4})}{4!} \delta^3 u_{1/2} + \cdots\right]$

$$+ \left[\dfrac{q^2 - \frac{1}{4}}{2!} \delta u_{-1/2} + \dfrac{(q^2 - \frac{1}{4})(q^2 - \frac{9}{4})}{4!} \delta^3 u_{-1/2} + \cdots\right].$$

Formula (3.19) is the most useful of the Everett formulae; it has the outstanding feature that, if used in subtabulation, the quantity within each bracket can be used twice.

EXAMPLE 3.3. Use Everett's Formula to obtain the values of $u_x = 10^7(\log_{10} x - 3)$ from $x = 1030(1)1050$.

x	u_x	δu_x	$\delta^2 u_x$	$\delta^3 u_x$
1000	0			
		43214		
1010	43214		−426	
		42788		+8
1020	86002		−418	
		42370		+9
1030 ——	128372 ——		−409 ——	+8
		41961		+8
1040 ——	170333 ——		−401 ——	
		41560		+7
1050	211893		−394	
		41166		+7
1060	253059		−387	
		40779		
1070	293838			

For this example, (3.19) takes the form

$$u_x = x\left[u_1 + \frac{x^2 - 1}{6}\,\delta^2 u_1\right] + y\left[u_0 + \frac{y^2 - 1}{6}\,\delta^2 u_0\right].$$

Let us write

$$F(x, u_a) = x\left[u_a + \frac{x^2 - 1}{6}\,\delta^2 u_a\right],$$

and compute values of $F(x, u_a)$ in parallel columns (we keep two guard decimals). Then, for $0 \le x \le 1$, we have

$$u_x = F(x, u_1) + F(1 - x, u_0).$$

x	$F(x, 1030)$	$F(x, 1040)$	$F(x, 1050)$
.1	12842.75	17039.92	21195.80
.2	25687.49	34079.43	42391.21
.3	38530.21	51118.15	63585.83
.4	51371.70	68155.66	84779.26
.5	64211.56	85191.56	105971.13
.6	77049.38	102225.46	127161.02
.7	89884.74	119256.96	148348.54
.8	102717.23	136285.65	169533.31
.9	115546.46	153311.04	190714.93

We obtain

$$u_{1031} = F(.1, 1040) + F(.9, 1030) = 132586.4,$$

and the other values are found similarly. Indeed, for an extended subtabulation, it is convenient to write the alternate columns $F(x, 1040)$, $F(x, 1060)$, etc., in reverse order; the subtabulated values can then be found by horizontal addition. The complete table follows.

x	u_x	x	u_x
1030	128372	1040	170333
1031	132586	1041	174507
1032	136797	1042	178677
1033	141003	1043	182843
1034	145205	1044	187005
1035	149403	1045	191163
1036	153597	1046	195317
1037	157787	1047	199467
1038	161973	1048	203613
1039	166154	1049	207755

All values agree with the values given in a large table, except that the under-lined final digits are one unit too small.

EXERCISES

1. The function $H'(x) = \dfrac{2}{\sqrt{\pi}} e^{-x^2}$ is tabulated below.

x	$10^{15} H'(x)$		
.720	67191	88112	40195
.721	67095	12735	11554
.722	66998	37890	31807
.723	66901	63615	91317
.724	66804	89949	74631
.725	66708	16929	60468
.726	66611	44593	21688

Compute $H'(.7233)$ by using the Gauss Formulae, Stirling's Formula, Bessel's Formula, and Everett's Formula. In each case, make a diagram showing what differences enter into the computation (the tabulated value is .66872 61450 47767).

2. The function $H(x) = \dfrac{2}{\sqrt{\pi}} \displaystyle\int_0^x e^{-t^2}\, dt$ is tabulated below.

x	$10^{10} H(x)$	
.455	48007	89899
.456	48099	59438
.457	48191	20618
.458	48282	73429
.459	48374	17860
.460	48465	53900
.461	48556	81539

Use Everett's Formula to subtabulate from .457(.0002).459.

3. From the table of $H(x)$ given below, compute $H(.1732)$ by Stirling's, Bessel's, and Everett's formulae.

x	$10^{15}\, H(x)$		
.170	18999	24612	01809
.171	19108	85102	16706
.172	19218	41844	46993
.173	19327	94818	30230
.174	19437	44003	06191
.175	19546	89378	16876
.176	19656	30923	06530
.177	19765	68617	21645

4.

x	$Ei(x) = \displaystyle\int_{-\infty}^{x} \frac{e^u}{u}\, du$	
.015	−3.6074	32975
.016	−3.5418	86664
.017	−3.4802	53746
.018	−3.4220	86531
.019	−3.3670	10002
.020	−3.3147	06894

Compute $Ei(.0173)$. The tabular value is $-3.4624\ 58001$.

5.

x	$\cos x$	
.066	.9978	22790
.068	.9976	88891
.070	.9975	51000
.072	.9974	09120
.074	.9972	63249
.076	.9971	13390

Compute $\cos .0718$ (tabular value .9974 23487). Check by using the Taylor series

$$\cos x = 1 - \frac{x^2}{2!} + \frac{x^4}{4!} - \cdots$$

6.

x	$\sinh x$	
.880	.9980	58397
.885	1.0051	35109
.890	1.0122	36949
.895	1.0193	64095
.900	1.0265	16726
.905	1.0336	95019
.910	1.0408	99155
.915	1.0481	29313

Compute $\sinh .8966$ (tabular value 1.0216 50156).

7.

x	$\log x$		
16100	9.68657	45509	725544
16150	9.68967	53286	508026
16200	9.69276	65212	204754
16250	9.69584	81877	578835
16300	9.69892	03867	948537
16350	9.70198	31763	253995
16400	9.70503	66138	122898
16450	9.70808	07561	935180
16500	8.71111	56598	886720

Compute log 16375 (tabular value 9.70351 10606 034527).

8.

x	e^x			
.500	1.64872	12707	00128	147
.501	1.65037	08166	06319	214
.502	1.65202	20128	83464	418
.503	1.65367	48611	82760	175
.504	1.65532	93631	57054	920
.505	1.65698	55204	60850	766
.506	1.65864	33347	50305	156
.507	1.66030	28076	83232	516
.508	1.66196	39409	19105	918

Compute $e^{.5043}$ (tabular value 1.65582 60364 63272 919).

9. Let

$$2B^{2n}(x) = \binom{x + n - 1}{2n} \quad \text{and} \quad B^{2n+1}(x) = \frac{2(x - \frac{1}{2})}{2n + 1} B^{2n}(x)$$

be the coefficients in Bessel's Formula (3.15). Let

$$E_0^{2n}(x) = -\binom{x + n - 1}{2n + 1}, \quad E_1^{2n}(x) = \binom{x + n}{2n + 1}$$

be the coefficients in Everett's Formula (3.18). Prove

(a) $E_1^{2n}(x) = E_0^{2n}(1 - x)$,

(b) $E_0^{2n}(x) = 2\dfrac{n + 1 - x}{2n + 1} B^{2n}(x)$,

(c) $E_1^{2n}(x) + E_0^{2n}(x) = 2B^{2n}(x)$,

(d) $E_1^{2n}(x) - E_0^{2n}(x) = 2B^{2n+1}(x)$.

10. Using the notation of Problem 9, prove that the two terms

$$E_0^{2n}(x)\, \delta^{2n}u_0 + E_1^{2n}(x)\, \delta^{2n}u_1$$

in Everett's Formula are equal to the two terms

$$2B^{2n}(x)\mu\delta^{2n}u_{1/2} + B^{2n+1}(x)\, \delta^{2n+1}u_{1/2}$$

in Bessel's Formula. (Hint: use Exercise 4 of Section 1.)

11. Show that the terms in $\delta^2 u_1$ and $\delta^4 u_1$ in Everett's Formula can be combined to give

$$\binom{x+1}{3}[\delta^2 u_1 - C\,\delta^4 u_1],$$

where C is a function of x such that $.15 \leq C \leq .20$ in the range $0 \leq x \leq 1$. Supposing that C is taken equal to a constant K, prove the following results.

(a) The error in so doing is given by

$$g(x) = \tfrac{1}{6}x(x^2 - 1)[K + \tfrac{1}{2}(x^2 - 4)].$$

(b) $g(x)$ has two extrema x_1 and x_2 in $0 \leq x \leq 1$.

(c) $g(x)$ has smallest maximal variation from zero for K a root of the equation

$$57600K^4 - 36800K^3 + 8640K^2 - 860K + 29 = 0.$$

(d) Taking $K = .181$, the value of $g(x)$ in (a) is $.00067$.

(e) If second differences $\delta^2 u_x$ are replaced by "modified second differences"

$$\delta^2_m u_x = \delta^2 u_x - K\,\delta^4 u_x,$$

then the error introduced is less than $\tfrac{1}{4}$ unit in the last figure, provided $\delta^4 u_x < 400$ (approximately).

5. THE ERROR IN INTERPOLATION FORMULAE

During the past three chapters, we have employed interpolation formulae freely with only the most general sort of consideration (Section 1.3) of the pitfalls which can be encountered. The general set-up can be briefly described as follows: we are given n values of a function $f(x)$ at the places

$$x_1 < x_2 < x_3 < \ldots < x_n;$$

we form a difference table and use it to interpolate, that is, we really replace $f(x)$ by a polynomial $p(x)$ with the property that

$$f(x_1) = p(x_1), \quad f(x_2) = p(x_2), \ldots, \quad f(x_n) = p(x_n).$$

If X is an arbitrary value in the interval $x_1 < X < x_n$, then the interpolation procedure produces the value $p(X)$; when we use this value $p(X)$ for $f(X)$, we are assuming (or hoping) that the identity of $f(x)$ and $p(x)$ at the places $x_1, x_2, \ldots, x_n$, ties $f(x)$ down so closely that it will not deviate very much from $p(x)$ in the whole interval $x_1 < x < x_n$. In our various examples, we have obtained an intuitive feeling that this desirable condition is likely to occur if $f(x)$ is a fairly smooth function having $x_1, x_2, \ldots, x_n$ fairly closely spaced, or if the sequence of differences $\Delta f, \Delta^2 f, \ldots$ decreases fairly rapidly. To put these considerations in an exact form, we prove

THEOREM 3.1. If $x_1 < x_2 < ... < x_n$, if $f(x)$ is a function with n continuous derivatives in the interval $x_1 < x < x_n$, and if $p(x)$ is the polynomial of degree $n-1$ such that

$$p(x_1) = f(x_1), \quad p(x_2) = f(x_2), \quad ... , \quad p(x_n) = f(x_n),$$

then

$$f(X) - p(X) = (X - x_1)(X - x_2) ... (X - x_n)\frac{f^{(n)}(\theta)}{n!},$$

where X is an arbitrary value in $x_1 < X < x_n$, and $\theta = \theta(X)$ lies in the range $x_1 \leq \theta \leq x_n$.

Proof. We select an arbitrary value X in the interval $x_1 < X < x_n$; in analogy with the method of proof used for the extended Theorem of Mean Value, we construct an artificial function

$$g(x) = f(x) - p(x) - \frac{f(X) - p(X)}{(X - x_1) ... (X - x_n)}(x - x_1)(x - x_2) ... (x - x_n).$$

It is clear that

$$g(x_1) = g(x_2) = ... = g(x_n) = g(X) = 0,$$

that is, $g(x)$ vanishes $n+1$ times in the interval $x_1 \leq x \leq x_n$; consequently, $g'(x)$ must vanish n times, $g''(x)$ must vanish $n-1$ times, etc., in the interval $x_1 \leq x \leq x_n$. In particular, $g^{(n)}(x)$ must vanish once in the interval, that is, there exists a number θ in $x_1 \leq \theta \leq x_n$ such that $g^{(n)}(\theta) = 0$. However,

$$g^{(n)}(x) = f^{(n)}(x) - 0 - \frac{f(X) - p(X)}{(X - x_1) ... (X - x_n)}n!;$$

putting $x = \theta$ in this expression, we obtain

$$0 = f^{(n)}(\theta) - \frac{f(X) - p(X)}{(X - x_1) ... (X - x_n)}n!.$$

Solving, we obtain

(3.22) $$f(X) - p(X) = (X - x_1)(X - x_2) ... (X - x_n)\frac{f^{(n)}(\theta)}{n!}.$$

Formula (3.22) is our desired result; it gives an expression for the difference between the true functional value $f(X)$ and the value $p(X)$ obtained by interpolation, for any point X in the interval from x_1 to x_n.

EXAMPLE 3.4. In Section 1.3, Exercise 3, the values of sin x were given at 15° intervals from 0° to 90°. What is the maximum error in interpolating to find sin X for any X within the range of the table?

In this problem, $f(x) = \sin x$, $n = 7$; then

$$f^{(7)}(x) = -\cos x.$$

It is obvious that (3.22) will not give us the maximum error, since we know nothing about θ except that it lies in the range $0°-90°$. However, since cos x is bounded in that interval, (3.22) will give us an upper bound on the size of the error, namely,

$$|f(X) - p(X)| \leq \frac{1}{7!}\left|(X - 0)\left(X - \frac{\pi}{12}\right)\left(X - \frac{2\pi}{12}\right) \cdots \left(X - \frac{6\pi}{12}\right)\right|.$$

In the earlier exercise, the value of sin $5\pi/24$ was computed; for $X = 5\pi/24$, the error is

$$\left|f\left(\frac{5\pi}{24}\right) - p\left(\frac{5\pi}{24}\right)\right| \leq \frac{1}{5040} 5 \cdot 3 \cdot 1 \cdot 1 \cdot 3 \cdot 5 \cdot 7 \frac{\pi^7}{24^7}$$

$$= 2.06 \times 10^{-7}.$$

This computation shows that any error in the computed value of sin $37°30'$ must arise from round-off; the error in using an interpolation polynomial to represent the function sin x will not affect the first five decimals of the answer.

Example 3.4 points up the chief drawback of Formula (3.22); a knowledge of the nth derivative $f^{(n)}(x)$ is required. Usually, we do not know the analytic expression for $f(x)$; we only know that it is a function with the given tabular values which probably runs along moderately smoothly between these values. The question then arises as to what we can do about estimating the error $f(X) - p(X)$ in the case when we do not have information concerning $f^{(n)}x$. Fortunately, the answer to this question is relatively easy; we can do nothing at all. It is this fact that $f^{(n)}(x)$ is rarely available in practical work which deprives Formula (3.22) of most of its usefulness.

One procedure which will sometimes give us an intuitive idea about the size of $f^{(n)}(x)$ is to note that

$$Df(x) \doteq \frac{f(x + h) - f(x)}{h} = \frac{1}{h}\Delta f(x).$$

Operationally, this equation can be written as

(3.23) $hD \doteq \Delta.$

Then

$$h^n D^n \doteq \Delta^n.$$

Consequently,

(3.24) $f^{(n)}(x) \doteq h^{-n}\Delta^n f(x).$

If the value of $\Delta^n f(x)$ does not change very rapidly in the interval, then the value of $D^n f(x)$ probably does not change very rapidly either, and a rough idea of the size of $D^{(n)}(\theta)$ can be gained by using (3.24) and the value of the nth difference.

EXAMPLE 3.5. Use (3.24) on the function $f(x)$ given in the following table.

x	$f(x)$	$\Delta f(x)$	$\Delta^2 f(x)$	$\Delta^3 f(x)$
.50	.52110			
		5705		
.55	.57815		145	
		5850		15
.60	.63665		160	
		6010		13
.65	.69675		173	
		6183		18
.70	.75858		191	
		6374		14
.75	.82232		205	
		6579		17
.80	.88811		222	
		6801		17
.85	.95612		239	
		7040		17
.90	1.02652		256	
		7296		20
.95	1.09948		276	
		7572		
1.00	1.17520			

The third difference is essentially constant; fourth differences would merely fluctuate irregularly around zero, and would only reflect round-off errors or experimental errors. The average value of the third difference is .00016; consequently, we can obtain an approximation to the third derivative $f'''(\theta)$, for θ between .50 and 1.00, as $.00016/(.05)^3 = 1.28$.

Actually, in this example, $f(x) = \sinh x$, $f'''(x) = \cosh x$, and the third derivative really ranges from 1.13 to 1.54 in the interval. We see that (3.24) has given a reasonable average value for the third derivative within the range .50–1.00.

EXERCISES

1. Consider the function $h(x) = x^3 - x$. Illustrate the fact that $h(x)$, $h'(x)$, and $h''(x)$ have 3, 2, and 1 zeros, respectively, in the interval $-1 \leq x \leq 1$, by plotting these three functions on the same sheet of graph paper (use a large scale). Such a graph is called a *Rolle diagram*.

2. Repeat Exercise 1 using the function $h(x) = x^4 - 3x^3 - x^2 + 3x$.

3. Discuss the error involved in approximating to the function e^x by
   (a) a polynomial of degree 4 determined by the values of e^x at the places 1.00, 1.01, 1.02, 1.03, 1.04; $X = 1.015$;
   (b) a polynomial of degree 4 determined by the values of e^x at the places 1.00, 1.10, 1.20, 1.30, 1.40; $X = 1.15$;

   (c) a polynomial of degree 4 determined by the values of e^x at the places 1.0, 1.5, 2.0, 2.5, 3.0; $X = 1.75$;

   (d) a polynomial of degree four determined by the values of e^x at the places 1, 2, 3, 4, 5; $X = 2.5$.

4. Compare the error involved in 3(b) with the error involved by

   (a) using a cubic polynomial which coincides with e^x for the x-values 1.00, 1.10, 1.20, 1.30;

   (b) using a quintic polynomial which coincides with e^x at the x-values 1.00, 1.10, 1.20, 1.30, 1.40, 1.50.

   In both cases, take $X = 1.15$.

5. In 4(a), find a bound on the maximum possible error for any point X in the range 1.00–1.30 (there is, of course, no guarantee that this bound is ever reached).

6. Employ the following table to establish that $f^{iv}(x)$ is probably less than 10^{-6}.

x	$f(x)$
50	39.19611753
51	39.79813617
52	40.39419423
53	40.98435072
54	41.56866408
55	42.14719216
56	42.71999224

7. Use Bessel's formula (3.17) to predict $f(112.5)$ by employing a cubic interpolating polynomial coinciding with $f(x)$ at the x-values 111, 112, 113, 114.

x	$f(x)$
110	413927
111	453230
112	492180
113	530784
114	569049
115	606978

   Show that the error involved should not exceed $\frac{1}{10}$ of a unit in the last digit [according to (3.24)].

8. Use (3.22) to obtain an error estimate in Exercise 7, given that $f(x) = 10^7(\log_{10} x - 2)$.

6. LEAST SQUARES APPROXIMATION

The word "interpolation" can be roughly explained as the "determination of intermediate values". Intelligent guessing is thus one possible form of interpolation. Up to this stage, we have employed polynomial interpolation, that is, we have been given functional values $f(x_1), \ldots, f(x_n)$ and have replaced the unknown function $f(x)$ by the polynomial $p(x)$, of degree at most $n - 1$, which is such that $p(x_i) = f(x_i)$ for $i = 1, 2, \ldots, n$. If the values $x_1, \ldots, x_n$ are finely enough spaced so that the maximal deviation between $f(x)$ and $p(x)$, that is, $|f(x) - p(x)|$, is very small in the interval $x_1 \leq x \leq x_n$, then the polynomial $p(x)$ provides excellent values for $f(x)$ in that interval. Indeed, if $f(x)$ is a tabulated function such as $\sin x$, then polynomial interpolation, assuming a sufficient number of values of $f(x)$ are given, is really the only satisfactory method of interpolation. Polynomial interpolation uses all the information given, since $p(x)$ actually passes *through* all of the points $[x_i, f(x_i)]$ for $i = 1, 2, \ldots, n$.

It would probably be useful to restrict the meaning of interpolation to "polynomial interpolation" and employ the term approximation for other cases. Let us do so in this section at least. Then there are cases where interpolation may not be the most efficacious procedure; these will arise when functional values y_i are obtained by measurement, and are thus merely approximations to the "true" values $F(x_i)$ given by the "true" function $F(x)$. In these cases, polynomial interpolation would be used if we knew the true values $F(x_i)$, for then we should wish to interpolate using a polynomial $p(x)$ which agreed as closely as possible with $F(x)$, that is, whose graph fitted up as snugly as possible to the graph of $F(x)$. But we are not given values $F(x)$; rather, we are given values y_i which follow the general trend of $F(x)$, but differ according to the expressions

$$\text{(3.25)} \qquad\qquad y_i = F(x_i) + \epsilon_i,$$

where ϵ_i represents the total error in the ith measurement (due to both random and systematic causes). Consequently, it would be pointless to use a polynomial interpolator which followed the values of y_i exactly, since such an interpolator would not only follow the trend of the values $F(x_i)$, but would also exactly reproduce all the errors ϵ_i. We should note, at this stage, that we assume that only the measurements y_i are subject to errors; the values x_i are fixed assigned arguments for which measurements y_i are made; if the values x_i are also subject to error, the problem is vastly more difficult, and will not be treated here.

EXAMPLE 3.6. It is known that a quantity $F(x)$ follows a quadratic law $F(x) = \alpha x^2$. The following *measurements* y_i of $F(x)$ are taken.

x_i	y_i
2	.973
4	3.839
6	8.641
8	15.987
10	23.794

Approximate the value of $F(x)$ for $x = 5$.

Here we could use polynomial interpolation for the data as they stand; our procedure (*not* good) would be to build a difference table, as follows.

x	y	δy	$\delta^2 y$	$\delta^3 y$	$\delta^4 y$
2	.973				
		2.866			
4	3.839		1.936		
		4.802		.608	
6	8.641		2.544		−2.691
		7.346		−2.083	
8	15.987		.461		
		7.807			
10	23.794				

We might then employ Bessel's formula, with origin at 4, to interpolate for y_5. Note what our assumptions would be—we should be assuming that the function $F(x)$ could be closely represented by a quartic curve, despite the facts that higher differences are not becoming small (and indeed are behaving quite erratically), *and* that we are given that the true function $F(x)$ is actually a quadratic function. What we need is a method whereby we can use the five given values of y to provide a good estimate a for the single parameter α; thence we could estimate F_5 as $25a$. The use of Bessel's formula and the difference table would give an estimate f_5 which mirrored all the observational errors in the five measurements; the quartic interpolator would have to wiggle around so much in order to exactly pass through the points representing the observed data that it would likely lose the trend of the data. And when the data are observations, we must try to preserve the trend rather than describe the errors at the expense of the trend.

What then are we to do, if we are given n observations y_i rather than n true values $F(x_i)$? The answer is that we must have some knowledge of the "true" curve $F(x)$ which the data y_i are supposedly following. We then select an estimate $f(x)$ for this true curve $F(x)$ in such a manner that the sum of squared deviations between the observed values y_i and their estimated values $f(x_i)$ be as small as possible.

In this section, we shall discuss in detail two cases, the very important case when it is known that the theoretical curve $F(x)$ is a straight line $\alpha + \beta x$, and the analogous case when it is known that the theoretical curve $F(x)$ is a second-degree curve $\alpha + \beta x + \gamma x^2$. Naturally, there are many other possibilities; in fact an entire book could be written on the important case when $F(x)$ is periodic and can be written as a sum of trigonometric functions.

EXAMPLE 3.7. Standard weights x are hung on a spring, and the corresponding lengths y are measured. It is known (Hooke's Law) that the lengths should follow a linear relation

$$F(x) = \alpha + \beta x.$$

Determine, according to the principle of least squares, estimates a and b for α and β.

Here we are given pairs of observed values $(x_1, y_1), (x_2, y_2), \ldots, (x_n, y_n)$. The theoretical values for $y_1, \ldots, y_n$ are $\alpha + \beta x_1, \alpha + \beta x_2, \ldots, \alpha + \beta x_n$, respectively. Hence we may form the sum of squared differences between observed and theoretical values as

$$S(\alpha, \beta) = \sum_{i=1}^{n}(y_i - \alpha - \beta x_i)^2.$$

$S(\alpha, \beta)$ is a function of α and β; consequently, if it is to be minimized, we require that

$$\frac{\partial S}{\partial \alpha} = \frac{\partial S}{\partial \beta} = 0.$$

The required estimates a and b will thus be the solutions of the two linear equations found by equating

$$\frac{\partial S}{\partial \alpha} = -2\sum_{i=1}^{n}(y_i - \alpha - \beta x_i)$$

and

$$\frac{\partial S}{\partial \beta} = -2\sum_{i=1}^{n}x_i(y_i - \alpha - \beta x_i)$$

to zero. Note that the x_i and y_i are not variables, but are observed numbers. Simplifying the resulting equations, we end up with the result

(3.26)
$$\begin{cases} \sum_{i=1}^{n}y_i = na + b\sum_{i=1}^{n}x_i, \\ \sum_{i=1}^{n}x_iy_i = a\sum_{i=1}^{n}x_i + b\sum_{i=1}^{n}x_i^2. \end{cases}$$

This gives a line $f(x) = a + bx$ which *estimates* the true line $F(x) = \alpha + \beta x$. Note that this is *not* the true line $F(x) = \alpha + \beta x$; there is absolutely no way, using mathematics and not magic, whereby we can obtain theoretical constants α and β from physical observations (x_i, y_i); one can only obtain estimates of α and β. Thus, for any index i, we have three values: y_i, the observed value; $F(x_i) = \alpha + \beta x_i$, the theoretical (and unknown) value; and $f(x_i) = a + bx_i$, the least-squares estimate of $F(x_i)$.

EXAMPLE 3.8. Apply the results of Example 3.7 to the numerical data

x_i	y_i
2	7.32
4	8.24
6	9.20
8	10.19
10	11.01
12	12.05

We form the table

x_i	x_i^2	y_i	$x_i y_i$
2	4	7.32	14.64
4	16	8.24	32.96
6	36	9.20	55.20
8	64	10.19	81.52
10	100	11.01	110.10
12	144	12.05	144.60
Σ 42	364	58.01	439.02

Then Equations (3.26) become

$$6a + 42b = 58.01, \qquad 42a + 364b = 439.02,$$

and these are readily solved to yield $a = 6.3733333$, $b = .4707143$. Consequently, we have fitted a straight line to the given data as

$$f(x) = 6.3733333 + .4707143x,$$

and this line may reasonably be used as an estimator of the true value $F(x)$ for any x in the range $2 \leq x \leq 12$.

At this stage, we must carefully distinguish between the *problem of curve-fitting* and the *problem of regression*.† In Examples 3.7 and 3.8, we have simply been concerned with using the principle of least squares as a criterion whereby we might obtain a line $f(x) = a + bx$ which followed the trend of the given data, and, consequently, might be taken as an estimate of the true line $F(x)$. Other criteria would produce other estimates, and any reasonable estimate might serve; a free-hand graph drawn smoothly to represent the trend of the data points might serve very well.

If we assume that the errors ϵ_i in Formula (3.25) are distributed according to a normal probability distribution which has a variance independent of x_i, then one may assign *fiducial intervals* to α and β in Example 3.7. Thus, under these suppositions, we can make the statement "There is a 95% probability that α lies in the interval $a - T_\alpha < \alpha < a + T_\alpha$, and there is a 95% probability that β lies in the interval $b - T_\beta < \beta < b + T_\beta$". Here the numbers T_α and T_β are determined by statistical reasoning; we shall not discuss them, since the problems of statistical inference which arise are rather deep and require a full course in basic statistics for their comprehension; in particular, it requires considerable maturity to even decide whether the assumptions required for setting up fiducial limits are met. If this is the case, we call our problem a regression problem, and the line $f(x) = a + bx$ is called a

† Actually, it would be better to speak of a regression-type problem; the word regression should be restricted to cases when we deal with bivariate probability distributions. However, the methods used in true regression are the same as in the problem we deal with here, and it has become customary to extend the usage of the word.

regression line. But if these assumptions are not met, then we call our problem a problem in curve-fitting, and the fitted line $f(x) = a + bx$ gives us no exact information concerning the true line $F(x) = \alpha + \beta x$; we can only offer up a pious and hopeful prayer that it be not too different from $F(x) = \alpha + \beta x$. In the rest of this section, we shall consider only the problem of "curve-fitting by least squares", that is, *when given a set of observed data points, to find the estimating curve of prescribed form which minimizes the sum of squared y-deviations between observed and estimated values.*

EXAMPLE 3.9. Find the least-squares estimator for the true curve

$$F(x) = \alpha + \beta x + \gamma x^2,$$

given data (x_i, y_i) for $i = 1, 2, \ldots, n$.

For the ith observation, the y-deviation between observed and theoretical values is

$$y_i - \alpha - \beta x_i - \gamma x_i^2,$$

and the sum of squared deviations is

$$S = \sum_{i=1}^{n} (y_i - \alpha - \beta x_i - \gamma x_i^2)^2.$$

Then

$$\frac{\partial S}{\partial \alpha} = -2\sum_{i=1}^{n}(y_i - \alpha - \beta x_i - \gamma x_i^2),$$

$$\frac{\partial S}{\partial \beta} = -2\sum_{i=1}^{n}x_i(y_i - \alpha - \beta x_i - \gamma x_i^2),$$

$$\frac{\partial S}{\partial \gamma} = -2\sum_{i=1}^{n}x_i^2(y_i - \alpha - \beta x_i - \gamma x_i^2).$$

Equating these three partial derivatives to zero, we obtain a, b, and c (the estimates of α, β, γ) as solutions of the equations

(3.27)
$$\begin{cases} \Sigma y = na + b\Sigma x + c\Sigma x^2, \\ \Sigma xy = a\Sigma x + b\Sigma x^2 + c\Sigma x^3, \\ \Sigma x^2 y = a\Sigma x^2 + b\Sigma x^3 + c\Sigma x^4. \end{cases}$$

Here, for simplicity, we omit the range of summation; henceforth, we shall always do so.

EXAMPLE 3.10. Apply the results of Example 3.9 to the case

x	y
2	3.07
4	12.85
6	31.47
8	57.38
10	91.29

In this case, since there is an odd number of items, we greatly simplify matters by introducing $X = (x - 6)/2$. Then we form

X	y	Xy	X^2y
-2	3.07	-6.14	12.28
-1	12.85	-12.85	12.85
0	31.47	.00	.00
1	57.38	57.38	57.38
2	91.29	182.58	365.16
Σ 0	196.06	220.97	447.67

Also, $\Sigma X^2 = 10$, $\Sigma X^3 = 0$, $\Sigma X^4 = 34$, and Equations (3.27) give us

$$196.06 = 5a + 10c,$$
$$220.97 = 10b,$$
$$447.67 = 10a + 34c.$$

Thence
$$a = 31.276286,$$
$$b = 22.097,$$
$$c = 3.967857,$$

and our estimator is

$$f(X) = 31.276286 + 22.097X + 3.967857X^2.$$

In terms of the original variables, this becomes

$$f(x) = .695999 - .855071x + .991964x^2,$$

and can now be used for estimating values. In particular, note that $y_2 = 3.07$, $f_2 = 2.953713$, whereas the true value F_2 is not known. Note also that the sum $\sum_{i=1}^{5} (y_i - f_i)^2$ is less for the estimator f_i found than for any other estimator.

EXAMPLE 3.11. Find the least-squares estimate for α, given n observations (x_i, y_i) and the prescribed curve $F(x) = \alpha x^n$, where n is a fixed number.

In this problem, we have only one parameter α; we readily find

$$S(\alpha) = \Sigma (y_i - \alpha x_i^n)^2,$$
$$\frac{\partial S}{\partial \alpha} = -2 \Sigma x_i^n (y_i - \alpha x_i^n).$$

Equating $\partial S / \partial \alpha$ to zero, we obtain the least-squares estimate a for α as

$$a = \frac{\Sigma x_i^n y_i}{\Sigma x_i^{2n}}.$$

This example is a good place to point out that other reasonable estimates of α exist, and will give slightly different approximating curves $f(x) = ax^n$. For instance, one might compute y_i/x_i^n for each of the n items of data, and then compute

$$a_1 = \frac{1}{n} \Sigma \frac{y_i}{x_i^n} ;$$

such an estimate a_1 is not, of course, the least-squares estimate a.

A different estimator a_2 can be found by taking the logarithms of the given data. Instead of a theoretical curve

$$F(x) = \alpha x^n,$$

consider

$$\log F(x) = \log \alpha + n \log x.$$

This is a linear relationship, and so may be written

$$F^*(x) = \alpha^* + n x^*,$$

where the asterisk denotes a logarithm. Our data then are taken in the form (x_i^*, y_i^*), where we keep an adequate number of decimals in order that the process of taking logarithms introduce no new errors. The least-squares estimate of α^* is immediately found to be

$$a^* = \frac{1}{n} [\Sigma y_i^* - n \Sigma x_i^*],$$

and our estimator a_2 of α is found from the relation

$$\log a_2 = a^*.$$

While this is a least-squares estimate, $a_2 \neq a$. We found a by fitting a curve $F(x) = \alpha x^n$ to the given data, whereas a_2 was found by fitting a straight line to the logarithms of the items of data; the two processes are not identical.

EXAMPLE 3.12. Apply the three methods of estimation of Example 3.11 to the data of Example 3.6.

Here we can not change the origin without altering the form of the curve $F(x) = \alpha x^2$; however, it is easy to verify that

$$\Sigma x_i^4 = 15664, \qquad \Sigma x_i^2 y_i = 3778.96,$$

and consequently

$$a = .24125.$$

The estimate $a_1 = \frac{1}{5}\Sigma y_i/x_i^2$ is found as

$$a_1 = .24219.$$

The estimate a_2 requires a table.

x_i^*	y_i^*
.30103	$\bar{1}$.98811
.60206	.58422
.77815	.93656
.90309	1.20377
1.00000	1.37647

Here we use logarithms to base 10, although any base would serve. Then

$$\Sigma\, y_i^* = 4.08913, \qquad \Sigma\, x_i^* = 3.58433,$$
$$a^* = -.61591 = \bar{1}.38409,$$
$$a_2 = .24214.$$

From the point of view of curve-fitting, there is little to choose among these estimates a, a_1, a_2.

It would be well to conclude this section by working out an example involving a numerical computation where students frequently go wrong. This occurs in computing the minimal sum of squared deviations between estimated and observed values, namely,

$$m = \Sigma\, [y_i - f(x_i)]^2$$

in the case of a linear curve $F(x) = \alpha + \beta x$; this quantity m is very important in regression problems, when one can reasonably make further assumptions about the data and needs to set up fiducial intervals for α and β.

EXAMPLE 3.13. Compute the minimal sum of squared deviations in Example 3.8.

In Example 3.8, it is easy to make the table

x_i	y_i	$f(x_i)$	$y_i - f_i$
2	7.32	7.314762	$+.005238$
4	8.24	8.256191	$-.016191$
6	9.20	9.197619	$+.002381$
8	10.19	10.139048	$+.050952$
10	11.01	11.080476	$-.070476$
12	12.05	12.021905	$+.028095$

As a check, we should have

$$\Sigma\, (y_i - f_i) = \Sigma\, (y_i - a - bx_i) = 0;$$

actually, it comes to $-.000001$. Accumulating the sum of squares, we obtain

$$m = \Sigma\, (y_i - f_i)^2$$
$$= .00864756.$$

Now this evaluation involves little work, since $n = 6$; however, for larger n, it is advantageous to simply note that

$$m = \Sigma \, (y_i - a - bx_i)^2$$

$$= \Sigma \, (y_i - a - bx_i)y_i - a \Sigma \, (y_i - a - bx_i) - b \Sigma \, x_i(y_i - a - bx_i),$$

and both the second and third terms vanish, in virtue of the equations determining a and b; then

$$m = \Sigma \, y_i^2 - a \Sigma \, y_i - b \Sigma \, x_i y_i$$

$$= 576.3787 - 6.3733333(58.01) - 439.02(.4707143)$$

$$= .00864328.$$

Note that, while we used seven decimals in b, b was multiplied by 439.02; so the round-off error in the seventh digit of b moves up to the fifth decimal in m; this leaves us with only three-figure accuracy, $m = .00864$, despite the retention of seven figures in b. This illustrates the fact that it may often be necessary to retain ten or eleven figures in a and b in order to compute m; too often, students give their estimating line as

$$f(x) = 6.373 + .471x$$

(after all, they claim, that's one more decimal than in the data!); then they compute

$$m = \Sigma \, y_i^2 - a \Sigma \, y_i - b \Sigma \, x_i y_i$$

$$= -.09745,$$

thus obtaining m, *a sum of squares*, as a negative number, very different from the value .00864.

The way in which m is used is the following: if the errors ϵ_i in (3.25) follow the normal probability distribution, then the following quantities are computed:

(3.28)
$$\begin{cases} T_\alpha = t_{n-2}\sqrt{\dfrac{m}{n-2}\left[\dfrac{1}{n} + \dfrac{\bar{x}^2}{\Sigma \, (x_i - \bar{x})^2}\right]}, \\[4mm] T_\beta = t_{n-2}\sqrt{\dfrac{m}{(n-2)[\Sigma \, (x_i - \bar{x})^2]}}. \end{cases}$$

Here $\bar{x} = \frac{1}{n}\Sigma \, x_i$, and t_{n-2} is a numerical constant tabulated as "Student's t"; tables give t_4, for 95% fiducial limits, as 2.776. Using this value in the present example, we have

$$T_\alpha = .03978,$$

$$T_\beta = .010397.$$

Consequently, there is a 95% probability that β lie in the range

$$b - .010397 < \beta < b + .010397,$$

and that α lie in the range

$$a - .03978 < \alpha < a + .03978.$$

EXERCISES

1. Use the principle of least squares to obtain estimates of the parameters, given n observations (x_i, y_i), and being required to fit a curve of prescribed form, namely,

(a) $F(x) = \alpha/x$,
(b) $F(x) = \alpha + \beta x^3$,
(c) $F(x) = \alpha + \beta x + \gamma x^3$,
(d) $F(x) = \alpha + \beta x + \gamma x^2 + \delta x^3$,

(e) $F(x) = 1 + \beta x$,
(f) $F(x) = \alpha x$,
(g) $F(x) = \alpha e^{\beta x}$,
(h) $F(x) = \log_e (\alpha + \beta x)$.

2. Find the least-squares line fitting the following data.

x_i	2.5	3.0	3.5	4.0	4.5	5.0	5.5
y_i	4.32	4.83	5.27	5.74	6.26	6.79	7.23

Set up 95% fiducial intervals for α and β on the assumption that the requirements for a regression problem are met.

3. If the prescribed curve is $F(x) = \beta x$, estimate β by least squares, given

x_i	3	4	5	6	7
y_i	2.13	2.86	3.60	4.29	4.95

4. The following data are obtained in investigating a linear relation $F(x) = \alpha + \beta x$.

x_i	4	6	8	10	12
y_i	13.72	12.90	12.01	11.14	10.31

Obtain the least-squares estimate $f(x) = a + bx$ provided by these data. Set up 95% fiducial limits for α and β, assuming that the errors ϵ_i satisfy the requirements of a regression problem ($t_3 = 3.182$).

5. Repeat Exercise 4, using the data

x_i	3	3.5	4	4.5	5
y_i	9.84	9.55	9.29	8.96	8.68

6. You are given n pairs of observations (x_i, y_i), and are required to fit a curve of the form $y = \alpha/x^2$ to them. Find the best-fitting curve by
(a) direct application of the method of least squares;
(b) transformation of the equation into a linear relationship.
Fit a curve of the above form to the following data.

x_i	5	10	15	20	25	30
y_i	.02243	.00546	.00297	.00147	.00098	.00067

7. In the following table, y_i represents the amount of a substance which can be dissolved in a fixed amount of water at temperature x_i.

x_i	10	15	20	25	30	35
y_i	15.005	15.330	15.627	15.921	16.191	16.462

Fit a curve of the form

$$F(x) = \alpha + \beta x + \gamma x^2 + \delta x^3$$

to these data.

8. Fit a curve of the form $F(x) = \alpha e^{\beta x}$ to the following data by using a least-squares line upon x_i and log y_i.

x_i	10	15	20	25	30	35
y_i	.512	.708	.906	1.192	1.529	1.926

9. Repeat Exercise 8 for the data

x_i	5	6	7	8	9	10
y_i	.308	.274	.243	.224	.200	.181

10. Fit a curve

$$F(x) = \alpha + \beta x + \gamma x^2$$

to the data

x_i	y_i
2.0	7.06
2.2	11.34
2.4	15.62
2.6	19.50
2.8	25.62
3.0	31.94
3.2	37.02
3.4	44.32
3.6	51.56
3.8	58.72
4.0	67.08
4.2	75.91

11. Repeat Exercise 10 if the prescribed curve is

$$F(x) = \alpha + \beta x + \gamma x^2 + \delta x^3.$$

12. Compare the fitted values f_i in Exercises 10 and 11 by computing $f(x_i)$ for $x_i = 2.0(.2)4.2$.

13. Fit the curve $F(x) = \log (\alpha + \beta x)$ to the following data by converting the given relation to a linear form

$$e^{F(x)} = \alpha + \beta x$$

and using "data" e^{y_i} and x_i.

x_i	2	3	4	5	6	7
y_i	1.952	2.156	2.413	2.549	2.670	2.821

14. Apply the method of least squares, given a prescribed function of two variables

$$F(x, y) = \alpha + \beta x + \gamma y$$

and the n observed points (x_i, y_i, z_i).

15. Fit the curve $F(x) = 1/(\alpha + \beta x)$ to the following set of data by using the linear relation

$$[F(x)]^{-1} = \alpha + \beta x.$$

x_i	5	6	7	8	9
y_i	1.335	1.431	1.247	1.197	1.118

16. Fit the curve $F(x) = \alpha e^{\beta/x}$ by linearizing it as

$$\log F(x) = \log \alpha + \beta x^{-1},$$

where $\log F(x)$ and x^{-1} are linearly related.

x_i	10	20	30	40	50	60
y_i	60.74	14.82	8.25	6.37	5.49	4.97

Inverse Interpolation and the Solution of Equations

1. THE PROBLEM OF INVERSE INTERPOLATION

Suppose that we are given a table of functional values; then the problem of direct interpolation, which we have so far been considering, can be phrased briefly as "Given x, find $f(x)$". The problem of inverse interpolation takes the form "Given $f(x)$, find x".

It is important to realize that these two problems are, in general, distinct from one another. For example, let us tabulate (Table 4.1) the function

$$f(x) = 5x^3 - 2.5x^2 + 1.4x - 1.8.$$

Table 4.1. DIFFERENCES OF THE FUNCTION
$5x^3 - 2.5x^2 + 1.4x - 1.8$

x	$f(x)$	$\delta f(x)$	$\delta^2 f(x)$	$\delta^3 f(x)$
−2	−54.6			
		43.9		
−1	−10.7		−35.0	
		8.9		30.0
0	−1.8		−5.0	
		3.9		30.0
1	2.1		25.0	
		28.9		
2	31.0			

Then we may immediately find, using Bessel's formula, that

$$f(\tfrac{1}{2}) = \tfrac{1}{2}(2.1 - 1.8) - \tfrac{1}{16}(-5.0 + 25.0) = -1.1.$$

Of course, in general, we will not know that $f(x)$ is indeed a polynomial; rather we shall be assuming that the values of x are finely enough spaced so that $f(x)$ is accurately approximated by a polynomial within the limits of the table.

Now suppose we consider $f(x)$ as independent variable, $x = g(f)$ as dependent variable; then we may form a table of (necessarily) divided differences of x. The result is given in Table 4.2.

Table 4.2

f	$x = g(f)$	$10\Delta g$	$10^2\Delta^2 g$	$10^3\Delta^3 g$	$10^4\Delta^4 g$
−54.6	−2				
		.22779			
−10.7	−1		.16966		
		1.12360		.16856	
−1.8	0		1.12539		−.070165
		2.56410		−.43205	
2.1	1		−.67624		
		.34602			
31.0	2				

We may then apparently write

$$g(f) = g(-1.8) + (x + 1.8)\Delta g + (x + 1.8)(x - 2.1)\Delta^2 g$$
$$+ (x + 1.8)(x - 2.1)(x + 10.7)\Delta^3 g$$
$$+ (x + 1.8)(x - 2.1)(x + 10.7)(x - 31.0)\Delta^4 g,$$

where the divided differences used are those on the solid line in Table 4.2. It is then easy to compute

$$g(-1.1) = .179487 - .025209 + .009291 - .004843 = .15873.$$

The same result would, of course, be obtained using the Lagrange Interpolation Formula, although the labour involved would be somewhat greater.

This example shows that straight interchange of the two variables to give x as a function of f, and subsequent interpolation for x, given f, yields a very incorrect result (.159 as opposed to the correct value .5). Wherein lies the difficulty, and why is the procedure illegitimate?

To understand the answer to this question, we must examine the assumptions underlying the direct interpolation of Table 4.1 and the (faulty) method of inverse interpolation used in Table 4.2. When we perform a direct interpolation, we assume either that $f(x)$ *is* a polynomial (as it is in Table 4.1) or that it can be closely *represented* by a polynomial. But the fact that $f(x)$ is a polynomial (or very closely approximable by one) does not guarantee that the inverse function $g(f)$ will be a polynomial or even very closely represented by one. Tables 4.1 and 4.2 show how, even when third differences of $f(x)$ are constant, fifth differences of $g(f)$ can be very appreciable; $g(f)$ could only be represented accurately by a polynomial of higher degree, and insufficient values and differences are available to permit such a representation.

There is one case when inverse interpolation can be performed by the method of Table 4.2; that is when the differences of the function g become approximately constant, thus indicating that g, like f, can be closely represented by a polynomial. In other words, the spacing of x-values *and* the spacing of f-values are *both* sufficiently fine to permit polynomial representation of $f(x)$ and of $g(f)$. This case, while not of common occurrence, does crop up for some very finely tabulated functions.

We are thus left squarely confronted by the problem: since the procedure of Table 4.2 is, in general, inadmissible, how should we solve the inverse interpolation problem "Given $f(x)$, find x"? Various procedures for solution of this problem will be discussed in the succeeding sections.

EXERCISES

1. Confirm that the Lagrange interpolation formula, applied to Table 4.2, gives $g(-1.1) = .15873$.

2. Let the following values of $y = f(x)$ be given.

x	y
0	.8
1	5.3
2	21.8
3	50.3

Graph accurately the (quadratic) polynomial $y = f(x)$ representing these data, and use this polynomial to find $f(2.3) = a$. Then graph accurately the (cubic) polynomial $x = g(y)$ given by the Lagrange formula; find $g(a)$ and note how $g(a)$ differs from 2.3.

3. The following table of sin $x°$ is given.

$x°$	sin $x°$
24	.40674
25	.42262
26	.43837
27	.45399
28	.46947
29	.48481

Show that $\sin 26°31' = .44620$; then use the Lagrange interpolation formula to solve the problem "Given $\sin x° = .44620$, find x". (Note that the result here is good because of the fine spacing in both argument and entry.)

4. Let $y = x^3$ be given for $x = 0, 1, 2, 3, 4, 5$. Try the Lagrange formula to obtain x, given $y = 3.375$. Compare this result with the correct value 1.5.

5. The function $y = x^2 - 4x + 9$ is given for $x = 0, 1, 2, 3, 4, 5$. What value of x makes y equal to 8? How does this value compare with the result of making y the independent variable and employing the Lagrange formula?

2. CONNECTION BETWEEN INVERSE INTERPOLATION AND THE SOLUTION OF EQUATIONS

The problem of solving an equation in one unknown x can be simply formulated as "Given

$$(4.1) \qquad\qquad f(x) = 0,$$

find all possible values of x". The problem of inverse interpolation takes the similar form "Given $f(x) = a$, find the value of x corresponding to this given value of $f(x)$". By transposing the a to the left-hand side of the equation, or, alternatively, decreasing all functional values by a, this problem can be reduced to the form of (4.1) "Given $f(x) = 0$, find the value of x corresponding to the value $f(x) = 0$". We thus see that inverse interpolation is a special case of (4.1); we need to solve an equation, but we know that

1) the root x we want is real;
2) the function $f(x)$ is (or is representable as) a polynomial of relatively low degree;
3) the root x is isolated for us (that is, we know, before starting, the approximate location of our answer).

Consequently, if we study the general problem of solving Equation (4.1), we shall cover that particular case of (4.1) when we want a real root, whose value is roughly known, for a polynomial equation. Indeed, we shall often solve (4.1) by a method of inverse interpolation, and this will necessitate methods for approximately locating real roots; the most potent is the simple graphic approach described in the next section.

3. GRAPHIC SOLUTION OF EQUATIONS

Frequently Equation (4.1) can usefully be written in the form $f_1(x) = f_2(x)$. The solution will then be provided by the intersections of the two curves $y = f_1(x)$ and $y = f_2(x)$.

EXAMPLE 4.1. The equation $e^{at} - at - b = 0$ occurs in heat transfer. Solve, given $a = .4, b = 9$.

Here we write $e^{at} = at + b$ and graph the two standard curves $y = e^{at}$

and $y = at + b$ (Figure 4.1). From the graph, it is immediately clear that there are two roots, a negative root nearly equal to the value $-9/.4 = -22.5$, and a positive root (the physically significant one) which is a little greater than 6. (For $t = 6$, the values of $e^{.4t}$ and $.4t + 9$ are 11.023 and 11.4 respectively.)

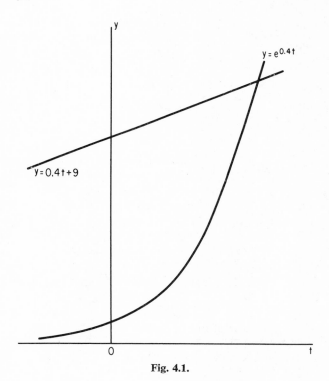

Fig. 4.1.

EXAMPLE 4.2. The equation $\tan x - kx = 0$ occurs in mechanical engineering; solve this equation for $k = 1.5$, .5, and $-.5$.

Here we write $\tan x = kx$, and consider the intersections of the standard curves $y = \tan x$ and $y = kx$ (Figure 4.2). Clearly, in addition to the root at $x = 0$, we can read off approximate values for all the other roots; indeed, except for the first few roots (when arranged according to absolute values), the roots are very close to being equal to $(2n + 1)\pi/2$, where n assumes integral values.

EXAMPLE 4.3.
$$x^4 - 2x^3 + x^2 - 5 = 0.$$
Writing this equation as the intersection of $y = x^3(x - 2)$ with the

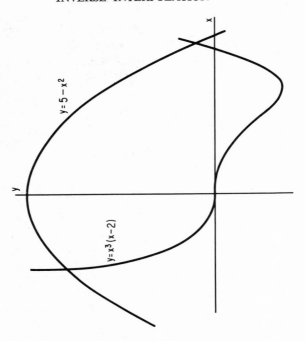

Fig. 4.3.

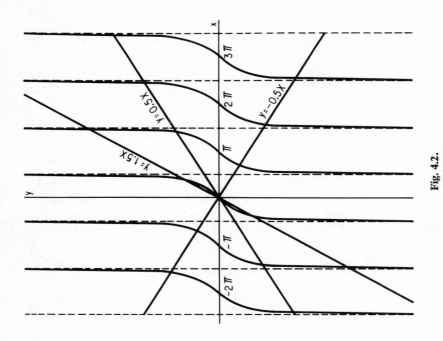

Fig. 4.2.

parabola $y = 5 - x^2$, we see there are two roots (Figure 4.3). The roots are roughly 2^+ and -1^-.

This example illustrates the fact that quite often more than one choice is available to us when we write our equation in the form $f_1(x) = f_2(x)$. Here it would have been equally advantageous to write the equation as the intersection of $y = x^2(x - 1)^2$ with the horizontal line $y = 5$.

In the case of polynomials of high degree (especially), the graphic approach often needs to be supplemented by a finite-difference table. An excellent illustration of the problems which may confront one is furnished by

EXAMPLE 4.4.

$$f(x) = x^6 - 27x^5 + 105x^4 - 140x^3 + 81x^2 - 21x + 2 = 0.$$

First we compute the sequence

$$y_1 = x - 27, \quad y_2 = xy_1 + 105, \quad y_3 = xy_2 - 140, \quad y_4 = xy_3 + 81,$$

$$y_5 = xy_4 - 21, \quad f(x) = xy_5 + 2.$$

This computation is very readily effected; thus, for $x = 5$, we find -22, -5, -165, -744, -3741, -18703. The procedure employed in this computation is so useful that we introduce a new notation; a simple bracket] will mean that the following symbol multiplies *everything* that precedes. Similarly, [will indicate that the preceding symbol multiplies everything that follows. With this notation,

$$f(x) = x - 27]x + 105]x - 140]x + 81]x - 21]x + 2.$$

Computing $f(x)$ for $x = -1(1)5$, we can check our work by finding $\delta^6 = 6! = 720$. The table can then be extended, using the fact that sixth differences are constant (and the structure of the table is quite informative). We note in addition that $f(x) > 0$ for x negative; also, writing

$$f(x) = x^5(x - 27) + 105x^3(x - \tfrac{140}{105}) + 81x^2(x - \tfrac{21}{81}) + 2,$$

we see that $f(x) > 0$ for $x \geq 27$. Hence all roots lie between $x = 0$ and $x = 27$. The complete tabulation is given in Table 4.3. Despite the size of the table, half an hour is ample to construct it; the construction should be checked periodically by computing certain values, say $f(12), f(18), f(24)$. Table 4.3 describes completely the behaviour of $f(x)$; there are two real roots at about 22.5 and 2^+; it is not obvious whether the curve crosses the x-axis between zero and one (for the answer, see Section 8 of this chapter).

Table 4.3

x	f(x)	Δ	Δ²	Δ³	Δ⁴	Δ⁵	Δ⁶
−1	377						
		−375					
0	2		374				
		−1		−330			
1	1		44		−240		
		43		−570		−2160	
2	44		−526		−2400		720
		−483		−2970		−1440	
3	−439		−3496		−3840		720
		−3979		−6810		−720	
4	−4418		−10306		−4560		720
		−14285		−11370		0	
5	−18703		−21676		−4560		720
		−35961		−15930		720	
6	−54664		−37606		−3840		720
		−73567		−19770		1440	
7	−128231		−57376		−2400		720
		−130943		−22170		2160	
8	−259174		−79546		−240		720
		−210489		−22410		2880	
9	−469663		−101956		2640		720
		−312445		−19770		3600	
10	−782108		−121726		6240		720
		−434171		−13530		4320	
11	−1216279		−135256		10560		720
		−569427		−2970		5040	
12	−1785706		−138226		15600		720
		−707653		12630		5760	
13	−2493359		−125596		21360		720
		−833249		33990		6480	
14	−3326608		−91606		27840		720
		−924855		61830		7200	
15	−4251463		−29776		35040		720
		−954631		96870		7920	
16	−5206094		67094		42960		720
		−887537		139830		8640	
17	−6093631		206924		51600		720
		−680613		191430		9360	
18	−6774244		398354		60960		720
		−282259		252390		10080	
19	−7056503		650744		71040		720
		368485		323430		10800	
20	−6688018		974174		81840		720
		1342659		405270		11520	
21	−5345359		1379444		93360		720
		2722103		498630		12240	
22	−2623256		1878074		105600		
		4600177		604230			
23	1976921		2482304				
		7082481					
24	9059402						

72

EXERCISES

1. Determine graphically (to one decimal) the roots of

$$x^3 + 3x^2 - 19x - 22 = 0.$$

2. In Example 4.2, let the large roots be written in the form

$$x_n = (2n + 1)\frac{\pi}{2} + \epsilon \qquad (n \text{ large}).$$

   Prove that
$$\epsilon \doteq \frac{-2}{\pi k(2n + 1)}.$$

3. Obtain the approximate values of the roots of the equation

$$8e^{-x^2} = x^2 + 2x - 2.$$

4. Determine the number (and approximate values) of real roots of the equations
   (a) $x^4 + 4x^2 + 5x + 20 = 0$,
   (b) $x^5 + 5x^4 + 18x^3 + 20x^2 + 20x - 60 = 0$,
   (c) $x^5 + 5x^4 - 8x^3 - 40x^2 + 8x + 30 = 0$.

5. Using the method of Example 4.4, discuss the equations
   (a) $x^6 - 20x^5 - 5x^4 + 90x^3 + 5x^2 - 70x - 1 = 0$,
   (b) $x^6 - 20x^5 + 80x^4 - 100x^3 + 70x^2 - 66x + 5 = 0$.

6. Discuss the nature of the roots of the equations
   (a) $\tan x = \log x$,
   (b) $x \sec x = 1$ (write as $\sec x = 1/x$),
   (c) $x \cos x = 1$,
   (d) $e^{-x/2} = 9 - x^2$,
   (e) $x^3 + x^5 = 2 \left(\text{write as } x^3 = \dfrac{2}{1 + x^2}\right)$,
   (f) $e^{-x} = x$.

7. Find from a large-scale graph approximate solutions of
   (a) $x^3 - 5x^2 - 6x + 1 = 0$,
   (b) $x^4 + 6x^3 - x^2 - 8x + 1 = 0$.

8. Discuss the equation

$$\sin x \cosh x - \cos x \sinh x = 0.$$

9. Draw the graph of the curves

$$y = x^2(x - 1)^2 \qquad \text{and} \quad y = 5$$

   (cf. Example 4.3).

4. ROOTS BY INVERSE INTERPOLATION

Suppose that a real root of $f(x) = 0$ has been isolated; then $f(x)$ may be tabulated, using a fine interval, in the neighbourhood of the root and represented closely by a central difference formula (exactly, if $f(x)$ should be a polynomial). This produces a polynomial equation in which the terms of higher degree have very small influence; consequently, an accurate approximation to the required root can be found, and this approximation can be improved by an iterative procedure.

EXAMPLE 4.5.

$$f(x) = x^3 + x^2 - 7x.$$

Find $x > 0$ so that $f(x) = -1$.

We reduce the problem to $x^3 + x^2 = 7x - 1$ and find, from Figure 4.4, that the required value is approximately 2. Tabulating $g(x) = f(x) + 1$ for values 1.9(.1)2.3, we obtain

x	X	$g(x)$	δg	$\delta^2 g$	$\delta^3 g$
1.9	-2	-1.831			
			.831		
2.0	-1	-1.000		.140	
			.971		.006
2.1	0	$-.029$		.146	
			1.117		.006
2.2	1	1.088		.152	
			1.269		
2.3	2	2.357			

We have thus reduced the given problem of solving $f(x) = -1$ to the inverse interpolation problem of finding x given $g(x) = 0$. Introducing $X = 10(x - 2.1)$, we find, using Stirling's formula,

$$g(X) = -.029 + \frac{X}{2}(2.088) + \frac{X^2}{2}(.146) + \frac{X(X^2 - 1)}{6}(.006).$$

For $g(X) = 0$, we have

$$1.044X = .029 - .073X^2 - .001X(X^2 - 1).$$

Dividing by 1.044, we obtain

$$X = .027778 - .069923X^2 - .000958X(X^2 - 1).$$

Clearly, $X_1 = .027778$ is a first approximation; using this value

$$X_2 = .027778 - .069923X_1^2 - .000958X_1(X_1^2 - 1)$$
$$= .027778 - .000054 + .000027 = .027751$$
$$X_3 = .027778 - .069923X_2^2 - .000958X_2(X_2^2 - 1)$$
$$= .027778 - .000054 + .000027 = .027751.$$

This process has converged rapidly to the root .027751 which, in the original

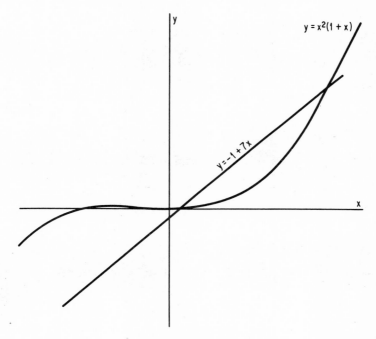

Fig. 4.4.

x-variable, is $2.1 + .0027751 = 2.1027751$. A greater number of decimals could have been secured in the answer by keeping more decimals in the recursion relation; as it is, we obtained seven decimals (the last dubious), and the approximations $X_1, X_2, X_3, \ldots$ can all be obtained mentally or on a slide rule.

Exercises 1 and 2 should prove instructive as to the merits of the various central difference formulae; in general, Stirling's formula is advantageous if the root lies in the range $|x| < \frac{1}{4}$; Bessel's formula is preferable if the root lies in the range $|x - \frac{1}{2}| < \frac{1}{4}$.

EXERCISES

1. Find the other two roots of $x^3 + x^2 - 7x + 1 = 0$, and check by finding their sum (which should equal -1).

2. Use the Bessel and Everett formulae on Example 4.5.

3. Justify the statements, made at the end of Section 4, concerning the relative merits of Stirling's and Bessel's formulae in inverse interpolation.

4. Find the real root of the equation

$$x^3 - 12x^2 + 124x - 1620 = 0.$$

Deduce the values of the complex roots.

5. Find the largest root of the equation

$$x^3 - 2x^2 - 9x + 3 = 0.$$

6. Solve completely the equation

$$x^4 - 6x^2 + 2x - 1 = 0.$$

7. Find the root between 1 and 2 of the equation

$$x^4 + x^3 + x^2 + 2x - 12 = 0.$$

5. ITERATIVE PROCEDURE FOR ROOTS

In the last section we employed a central difference formula to write the given equation in the form

(4.2) $$x = h(x).$$

We then found x as the limit of the sequence

(4.3) $$x_1, \quad x_2 = h(x_1), \quad x_3 = h(x_2), \quad x_4 = h(x_3), \quad \ldots$$

It is clear that if x_1 is selected so that the sequence $x_1, x_2, x_3, \ldots$ does converge, then the value obtained is a root of (4.2).

This procedure is a very important one, and we shall, in this section, illustrate other applications of it, where the equation (4.2) does not arise from a central difference formula. In short, if any equation can be written in the form (4.2), and if the sequence (4.3) converges to a value β, then β is a root of (4.2); the procedure used in obtaining sequence (4.3), whereby the answer at any stage is substituted in the given function to produce the next number in the sequence, is called an *iterative* procedure.

EXAMPLE 4.6. Solve the equation $x = 1 + \dfrac{x^4}{16}$. (Equations of this general form are frequent in applied probability; the root $x = 2$ is of no interest, but the root $x = 1^+$ is wanted in practice.)

Put $x_1 = 1$, $h(x) = 1 + (\tfrac{1}{2}x)^4$; we form successively

$$x_2 = h(x_1) = 1.0625,$$
$$x_3 = h(x_2) = 1 + .53125^4 = 1.07965,$$
$$x_4 = h(x_3) = 1 + .53983^4 = 1.08492,$$
$$x_5 = h(x_4) = 1 + .54246^4 = 1.08659,$$
$$x_6 = 1 + .54330^4 = 1.08713,$$
$$x_7 = 1 + .54357^4 = 1.08730,$$
$$x_8 = 1 + .54365^4 = 1.08735,$$
$$x_9 = 1 + .54368^4 = 1.08737,$$
$$x_{10} = 1 + .54369^4 = 1.08738.$$

Problems such as this one are characterized by rather slow convergence, and are best handled by a calculator. The process can be shortened by rounding to an even end digit at all stages, and "helping out" the approximation by moving along with it. One would be more likely to follow the sequence

$$x_2 = 1 + .5^4 = 1.06,$$
$$x_3 = 1 + .53^4 = 1.08,$$
$$x_4 = 1 + .54^4 = 1.086,$$
$$x_5 = 1 + .543^4 = 1.087,$$
$$x_6 = 1 + .5435^4 = 1.0874,$$
$$x_7 = 1 + .5437^4 = 1.08738,$$
$$x_8 = 1 + .54369^4 = 1.08738.$$

With this set-up, use of tables eliminates any computation until the x_6 stage (the result 1.087 is thus obtained immediately, and will frequently suffice in practice).

EXAMPLE 4.7. The Millikan oil-drop experiment for the determination of the ratio of the charge to the mass of an electron leads to the equation (Stokes' Law)

$$v = \frac{2}{9} g \frac{r^2}{n} (\mu - \mu_1)\left(1 + \frac{.000617}{pr}\right).$$

Find r, the radius of the oil-drop, for an experiment in which $g = 980$, $n = 1.832 \times 10^{-4}$, $\mu_1 = .0012$, $\mu = .9052$, $p = 72$, $v = .00480$.

Substituting the given values, we obtain

$$r^2 = \frac{9nv}{2g(\mu - \mu_1)}\left[1 + \frac{.000617}{pr}\right]^{-1}$$

$$= 4.4667 \times 10^{-9}\left[1 + \frac{.000617}{72r}\right]^{-1}.$$

$$r = 6.6833 \times 10^{-5}\left[1 + \frac{8.5694}{10^6 r}\right]^{-1/2}.$$

Clearly

$$r_1 = 6.6833 \times 10^{-5},$$

$$r_2 = 6.6833 \times 10^{-5}\left[1 + \frac{8.5694}{66.833}\right]^{-1/2} = \frac{6.6833 \times 10^{-5}}{1.06236}$$

$$= 6.2910 \times 10^{-5},$$

$$r_3 = 6.6833 \times 10^{-5}\left[1 + \frac{8.5694}{62.910}\right]^{-1/2} = \frac{6.6833 \times 10^{-5}}{1.06593}$$

$$= 6.2699 \times 10^{-5}.$$

One beauty of the iterative method is that minor errors correct themselves; an error in r_i (unless so gross that it destroys convergence) at worst slows up convergence (and may even speed it up if made judiciously). The final result is

$$r_4 = \frac{6.6833 \times 10^{-5}}{1.06615} = 6.2686 \times 10^{-5},$$

$$r_5 = \frac{6.6833 \times 10^{-5}}{1.06616} = 6.2686 \times 10^{-5}.$$

The last digit is uncertain.

It might be noted that, if only the result 6.27×10^{-5} were required, the whole operation could be carried out very rapidly by hand or on the slide rule.

EXAMPLE 4.8.

$$x^2 - 87x + 1 = 0.$$

This quadratic should be written $x = 87 - 1/x$. The large root is then computed from the sequence

$$x_1 = 87,$$

$$x_2 = 87 - \frac{1}{x_1} = 86.988506,$$

$$x_3 = 87 - \frac{1}{x_2} = 86.988504.$$

As a by-product of the method, the small root is obtained as $1/x_2 =$ .011495772 (since the product of the roots is the constant term in the equation, namely 1). This example illustrates the undesirability of the quadratic formula for the case of a very small root (since the quadratic formula expresses such a root as the difference of two nearly equal numbers).

The method of Example 4.8 is very potent for large roots; we shall employ it to find the largest root of the equation in Example 4.4.

EXAMPLE 4.9.

$$x^6 - 27x^5 + 105x^4 - 140x^3 + 81x^2 - 21x + 2 = 0.$$

(This equation originates with B. C. Carter, *Dynamic Forces in Aircraft Engines, J. Roy. Aero. Soc.* **31** (1927), p. 278; see also the discussion of the same problem in Frazer, Duncan, Collar, *Elementary Matrices*, p. 318.)

Write the equation as

$$x = 27 - \frac{105}{x} + \frac{140}{x^2} - \frac{81}{x^3} + \frac{21}{x^4} - \frac{2}{x^5}.$$

Clearly if we require only three decimals, the last two terms are negligible. Indeed, we start from $x = 27 - \dfrac{105}{x}$

to give

$$x_1 = 27 - \frac{105}{27} = 23.1,$$

$$x_2 = 27 - \frac{105}{23.1} = 22.4,$$

$$x_3 = 27 - \frac{105}{22.4} = 22.3.$$

Passing to

$$x = \left[\frac{140}{x} - 105 \right] \frac{1}{x} + 27,$$

we find

$$x_4 = 22.57$$

$$x_5 = 22.62.$$

Finally,

$$x = \left[\left(\frac{-81}{x} + 140 \right) \frac{1}{x} - 105 \right] \frac{1}{x} + 27,$$

and

$$x_6 = 22.625,$$

$$x_7 = 22.626.$$

This suffices for most purposes. However, to proceed further, we revert to our previous single-bracket notation and write

$$x = \frac{-2}{x} + 21 \Big]\frac{1}{x} - 81 \Big]\frac{1}{x} + 140 \Big]\frac{1}{x} - 105 \Big]\frac{1}{x} + 27;$$

using $x_7 = 22.626$, we rapidly obtain the result

$$x = 22.625853.$$

It might be instructive to indicate the results of the iteration beginning with this last value 22.625853 (Table 4.4).

Table 4.4

Operation	Computation		Result
			-2
$\div x, +21$	$-.08839446 + 21$	$=$	20.91160555
$\div x, -81$	$.92423501 - 81$	$=$	-80.07576499
$\div x, +140$	$-3.53912690 + 140$	$=$	136.46087310
$\div x, -105$	$6.0311924 - 105$	$=$	-98.9688076
$\div x, +27$	$-4.3741470 + 27$	$=$	22.6258530

We have thus determined the root to six decimals.

In concluding this section, we must point out that if, in Equation (4.2), we start with an arbitrary value x_1 and form

$$x_2 = h(x_1), \quad x_3 = h(x_2), \quad \ldots,$$

then there is no guarantee that the sequence obtained will converge. If it does converge, we obtain a root of the equation; if it does not converge, we discard it. In general, if x_1 is near a root, we can expect convergence, but this need not always be the case; sometimes a value just above a root will start a divergent sequence, whereas a value just less than the root will start a convergent sequence. While this presents a theoretical problem, it causes no trouble in practice, since it speedily becomes evident if a sequence is diverging. Various of the exercises illustrate different possibilities; in particular, Exercise 14 gives a simple sufficient condition for convergence. Furthermore, Exercise 11 illustrates the fact that one way of writing an equation in the form $x = h(x)$ may not produce a convergent sequence, while a second way of writing the equation may produce a convergent sequence.

EXERCISES

1. Solve by iteration the quadratics
   (a) $x^2 + 45x - 2 = 0$,
   (b) $2x^2 + 75x + 3 = 0$.

2. Find r in Example 4.7 if $\mu = .9273$, $\mu_1 = .0013$, $v = .00472$, $p = 75$, g and n are unchanged (obtain a three-figure result).

3. Solve the equations
   (a) $x = 1 + .025x^2 - .004x^3$,
   (b) $x^3 - 20x + 20 = 0$.

4. Find the largest roots of the equations in Exercise 5 of Section 3 (work to three decimals).

5. There is a root of Example 4.9 between 2 and 3. Try putting $x_1 = 3$ in the recursion relation, and see what sort of a sequence results.

6. An object will fall $16t^2$ feet in t seconds; the velocity of sound is 1089 feet per second, and a stone dropped from a bridge of height x is heard to hit the water 2.44 seconds later. Show that the equation for x can be written

$$x = 16\left[2.44 - \frac{x}{1089}\right]^2,$$

and solve for x (note the extreme cumbrousness of attempting to use the quadratic formula here).

7. Show that $\sin 20°$ satisfies the equation

$$x = \frac{\sqrt{3}}{6} + \frac{4}{3}x^3 = .28868 + \frac{4}{3}x^3,$$

and use this relation to find $\sin 20°$ to 5 decimals (*help* the sequence along, since it converges very slowly).

8. The iterative procedure outlined in this section is sometimes called the "staircase method". Justify the nomenclature by making a large scale graph of the functions $y = x$ and $y = 1 + (\frac{1}{2}x)^4$, used in Example 4.6, near the root 1.087; mark on this graph the successive approximation points $A(x_1, x_1)$, $B(x_1, x_2)$, $C(x_2, x_2)$, $D(x_2, x_3)$, $E(x_3, x_3)$, $F(x_3, x_4)$, ... , and draw in the staircase $ABCDEF...$ approaching the required solution (that is, the intersection of the two curves).

9. Draw the staircase for Exercise 8.

10. In the equation $x = \frac{1}{3} + \frac{1}{3}x^2$, the roots are found by the quadratic formula to be about 2.62 and .38. Starting with $x_1 = 0$, iterate as far as x_4; do the same for $x_1 = 1$ and for $x_1 = 2$. What happens if $x_1 = 3$? If $x_1 = R + \epsilon$, where R is the exact root $(3 + \sqrt{5})/2$ and $\epsilon > 0$? (Illustrate on a diagram.)

11. The equation $x^7 + 28x^4 - 480 = 0$ was given by W. B. Davies (*Educational Times*, 1867, p. 108), and has been used by many later authors as an illustration. Show that the sequence defined by

$$x_1 = 2, \qquad x = \frac{480}{x^6} - \frac{28}{x^2},$$

does not converge to a limit, but that the sequence

$$x_1 = 2, \qquad x = \left[\frac{480}{28 + x^3}\right]^{1/4}$$

does approach the positive root 1.9229. Use this latter sequence to obtain the root. Illustrate both sequences on (separate) diagrams.

12. Show that the sequence

$$x_1 = 1, \qquad x_{n+1} = \sqrt{1 + x_n}$$

does approach a limit by establishing
(a) $x_n < 2$ for all n (the sequence is bounded);
(b) $x_{n+1} - x_n > 0$ for all n (the sequence is increasing);
(c) the limit is $\frac{1}{2}(1 + \sqrt{5})$.
Hint: use induction in parts (a) and (b).

13. Use the method of Exercise 12 to establish the convergence of the sequences used in Examples 4.6 and 4.8, and in Exercises 3(b), 6, 8, and 11.

14. Show by a geometric argument that the sequence (4.3) obtained from Equation (4.2) converges to a root β if x_1 is taken close enough to β so that, in the neighbourhood of β, we have
(a) $h(x)$ continuous and monotone,
(b) $h'(x)$ continuous and monotone,
(c) $|h'(x)| < 1$.
Make two diagrams to illustrate the cases when $h'(x)$ lies between 0 and 1, and when $h'(x)$ lies between 0 and -1 (take $h(x) = 1 + .2x^3$ and $h(x) = 1 - .2x^3$ as your examples).

6. THE REGULA FALSI

One of the oldest methods of finding roots is the *regula falsi*. It requires a knowledge of the approximate location of the root and the computation of

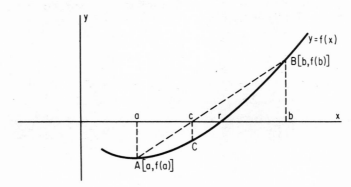

Fig. 4.5.

two values $f(a)$ and $f(b)$, where $a < r < b$, r being a root of $f(x) = 0$ (Figure 4.5). If a and b are close enough to r so that our function $f(x)$ is continuous in $a < x < b$, then $f(a)$ and $f(b)$ are opposite in sign. If we replace the arc AB by the chord AB, we obtain an abscissa c which is closer to

r than A was (in our figure $f(c)$ is negative). The value of c is obviously $[af(b) - bf(a)]/[f(b) - f(a)]$. The process may then be repeated using the chord BC.

Although there are various modifications of the *regula falsi* which cut down the amount of labour involved, it is clear that this process is slow and not particularly suited to hand computation. However, it is very simple and has many advantages in difficult problems which require an electronic computer. In such a case, simplicity of procedure is paramount, and the number of steps in the iteration is really of small import. The example is purely illustrative of the method; it can be solved more simply by other means.

EXAMPLE 4.10. Use the *regula falsi* on the equation

$$x^3 - 5x - 7 = 0,$$

noting that $\qquad f(2) = -9, \qquad f(3) = +5.$

Here $\qquad c_1 = \dfrac{2(5) - 3(-9)}{5 - (-9)} = \dfrac{37}{14} = 2.6.$

Then $\qquad f(2.6) = -2.424, \qquad f(3) = 5.$

Hence $\qquad c_2 = \dfrac{(2.6)5 + (2.424)3}{7.424} = \dfrac{20.272}{7.424} = 2.73.$

Then $\qquad f(2.73) = -.303583.$

Since we are getting close to the root, it is no longer suitable to use $f(3) = 5$; hence, we try for a better positive value and obtain $f(2.75) = .046875$. Then

$$c_3 = \frac{(2.73)(.046875) + (.303583)2.75}{.350458} = 2.7473.$$

$$f(c_3) = -.00082112.$$

The process can be continued as long as required; convergence can often be speeded up by using two values c_m and c_n having $f(c_m)$ and $f(c_n)$ of like sign. When c_m and c_n are both close to the root h, the chord joining $[c_m, f(c_m)]$ and $[c_n, f(c_n)]$ gives, when produced, a good value for r.

An even simpler procedure for slowly isolating a root is the Bolzano process of starting from a and b, as in Figure 4.5, and computing $f(c)$, where $c = \frac{1}{2}(a + b)$. Then the root r must lie in the interval (a, c) or in the interval (b, c). After n such interval bisections, r will be found as lying in an interval of length $2^{-n}(b - a)$. This procedure has great theoretical importance, but is not suited for hand computation.

EXAMPLE 4.11. Use the Bolzano bisection procedure on the equation

$$x^3 - 4x - 9 = 0.$$

Here $\qquad f(2) = -9, \qquad\qquad f(3) = +6.$

Then

$$f(2.5) = -3.375, \qquad f(3) = 6;$$
$$f(2.75) = +.797, \qquad f(2.5) = -3.375;$$
$$f(2.63) = -1.259, \qquad f(2.75) = .797;$$
$$f(2.69) = -.288, \qquad f(2.75) = .797;$$
$$f(2.72) = +.318, \qquad f(2.69) = -.288.$$

We have thus located the root as lying between 2.69 and 2.72; only an instrument with almost complete disregard for time can afford to use this method, though it is fundamentally of the utmost simplicity.

EXERCISES

1. Show that if, in Figure 4.5, a and b are such that $f(a)$ and $f(b)$ have like signs, then the formula

$$c = \frac{a f(b) - b f(a)}{f(b) - f(a)}$$

still holds.

2. Use the *regula falsi* to obtain one roots to three decimals of
   (a) $x^3 + 7x^2 + 9 = 0$,
   (b) $x^3 + x^2 + x + 7 = 0$.
   In each case, change an end-point whenever it becomes advantageous.

3. The equation

$$x^4 + 2x^2 - 16x + 5 = 0$$

has a root between 0 and 1. Carry out the Bolzano procedure on this equation four times.

7. NEWTON'S METHOD

Of all the methods advocated for the numerical solution of equations, Newton's method seems the one most generally satisfactory. Other methods, such as Horner's or Graeffe's, are restricted to polynomial equations and are of no advantage there; indeed, the Graeffe process is of such numerical complexity that texts which praise it extravagantly nevertheless usually restrict themselves to very simple examples.

Suppose that α is the desired root of the equation $f(x) = 0$; let x_1 be an abscissa near enough to α that the tangent at $P[x_1, f(x_1)]$ cuts the axis nearer to α than x_1. This point of intersection (Figure 4.6) is our second approximation x_2.

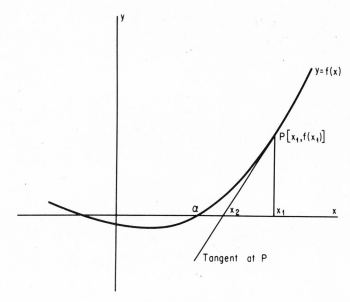

Fig. 4.6.

Since the tangent at P is

$$y - f(x_1) = f'(x_1)[x - x_1],$$

it is easy to find that

$$x_2 = x_1 - \frac{f(x_1)}{f'(x_1)},$$

Using x_2 as starting-point, the tangent at $[x_2, f(x_2)]$ will give

$$x_3 = x_2 - \frac{f(x_2)}{f'(x_2)}.$$

The process can be repeated, and the root α is approached with great rapidity. It is often very convenient to use the rule-of-thumb that, if the correction term $f(x_i)/f'(x_i)$ begins with n zeros after the decimal point, then the result is correct to about $2n$ decimals (that is, the number of correct decimals roughly doubles at each stage).

Another worthwhile feature of the Newton process is the fact that it is self-correcting for minor errors. Any errors made in determining x_2 will merely give a different point from which to draw the second tangent; this will not affect the limit α approached by the sequence $x_1, x_2, x_3, \dots$.

EXAMPLE 4.12. Find the smallest positive root of $x^3 - 5x + 3 = 0$.

Clearly there is a root between -2 and -3, a root between 1 and 2, and the required root between 0 and 1.

$$f(x) = x^3 - 5x + 3, \qquad f'(x) = 3x^2 - 5.$$

$$x_1 = 1, \qquad x_2 = 1 - \tfrac{1}{2} = .5,$$

$$x_3 = .5 + \frac{5}{34} = .5 + .14 = .64,$$

$$x_4 = .64 + \frac{.062144}{3.7712} = .64 + .0165 = .6565,$$

$$x_5 = .6565 + \frac{.000446412125}{3.70702325}$$

$$= .6565 + .000120 = .656620,$$

$$x_6 = .656620 + \frac{.000001597697528}{3.7065505268}$$

$$= .656620 + .000000431047$$

$$= .656620431047.$$

Convergence is thus very rapid, even though x_1 was not very near α.

EXAMPLE 4.13.
$$e^{.4t} = .4t + 9.$$

We have already discussed this transcendental equation graphically (Example 4.1), and found that there is a negative root at -22.5, a positive root at about 6. Consider

$$f(t) = e^{.4t} - .4t - 9,$$
$$f'(t) = .4(e^{.4t} - 1).$$

Then

$$t_1 = 6,$$

$$t_2 = 6 - \frac{-.37682}{4.009272} = 6.1,$$

$$t_3 = 6.1 - \frac{.03304}{4.189216} = 6.1 - .0079 = 6.0921.$$

This is as good a result as can be obtained using five decimal places in $e^{.4t}$.

EXAMPLE 4.14. Consider Example 4.9,

$$f(x) = x^6 - 27x^5 + 105x^4 - 140x^3 + 81x^2 - 21x + 2.$$

We shall determine the root situated at 2^+.

Using the bracket notation introduced previously, we have

$$f(x) = x - 27]x + 105]x - 140]x + 81]x - 21]x + 2,$$
$$f'(x) = 6x - 135]x + 420]x - 420]x + 162]x - 21.$$

If we try $x_1 = 2$, we obtain $f'(x_1) > 0$; hence we regress toward a smaller value from $x_1 = 2$, and must start at, say, $x_1 = 2.5$. We then have

$$f(2.5) = -22.765625, \qquad f(2.4) = 7.642496,$$

and the *regula falsi* suggests $x_2 = 2.43$. Then

$$f(2.43) = -.24457235, \qquad f'(2.43) = -279.64286,$$
$$x_3 = 2.43 - .000875 = 2.429125.$$

Since there are barely three zeros in the correction term, five decimals should be correct. Taking $x_4 = 2.429$, we obtain

$$f(2.429) = +.03450023, \qquad f'(2.429) = -278.5027152,$$

whence the root is

$$x_5 = 2.429 + .000124 = 2.429124.$$

Newton's method shows to even better advantage in Section 8, where it is employed to find the roots less than 1 of this same equation.

EXERCISES

1. Use Newton's method to find the three roots of $x^3 - 8x - 4 = 0$ to four places of decimals. Check your result by adding the roots.

2. Solve the equation
$$x^4 - x^3 - 8x^2 + x + 4 = 0$$
   to four places of decimals; check by adding the four roots.

3. The equation
$$x^3 + 2x^2 + 50x + 7 = 0$$
   has one real root a and two complex roots $b \pm ic$. Determine the value of a by Newton's method, and then find the two complex roots from the relations
$$a + (b + ic) + (b - ic) = a + 2b = -2,$$
$$a(b + ic)(b - ic) = a(b^2 + c^2) = -7.$$

4. Solve the equation $\tan x - 1.5x = 0$ of Example 4.2 to find the smallest positive root.

5. Solve the equation $x - 2 \sin 2x = 0$ (first locate the roots graphically).

6. A sphere of density d and radius r sinks in water to a depth x given by the equation

$$x^3 - 3rx^2 + 4r^3d = 0.$$

Solve this equation for $r = 1$, $d = .72$.

7. Solve completely the equation

$$x^5 + 2x^4 - x^3 - 20x^2 + x + 5 = 0.$$

8. Determine the other two roots in Example 4.12; check by finding their sum.

9. From the equation $x^2 - a = 0$, deduce the Newtonian iterative procedure

$$x_{n+1} = \frac{1}{2}\left(x_n + \frac{a}{x_n}\right)$$

for $\sqrt{a}$. Use this method to find $\sqrt{5} = 2.236067977$.

10. Apply Newton's method to the equation $1 - a/x^2 = 0$ to give the recursion relation

$$x_{n+1} = x_n\left[1 - \frac{x_n^2 - a}{2a}\right]$$

for $\sqrt{a}$.

11. Let a and $b \pm ic$ be the roots of the equation

$$x^3 + 3x^2 + 48 = 0.$$

Find these three roots by
(a) using the method of Exercise 3;
(b) dividing $x - a$ into $x^3 + 3x^2 + 48$ to obtain the quadratic factor $x^2 + px + q$.

12. Use the method of Exercise 9 to obtain the iterative formulae

$$x_{n+1} = \frac{1}{3}\left(2x_n + \frac{a}{x_n^2}\right),$$

$$x_{n+1} = \frac{1}{5}\left(4x_n + \frac{a}{x_n^4}\right),$$

$$x_{n+1} = \frac{1}{2}\left(x_n + \frac{1}{ax_n}\right),$$

for $\sqrt[3]{a}$, $\sqrt[5]{a}$, and $1/\sqrt{a}$, respectively.

13. Suppose that r_1, r_2, and r_3 are real roots of the cubic equation

$$x^3 + ax^2 + bx + c = 0,$$

and that they have been determined to n decimals, that is, each is in error by an amount $< \epsilon$, where $\epsilon < \frac{1}{2}(10^{-n})$. Use differentials to show that the maximal possible errors in

$$r_1 + r_2 + r_3, \qquad r_1 r_2 + r_1 r_3 + r_2 r_3, \qquad r_1 r_2 r_3,$$

are respectively

$$3\epsilon, \qquad 2(|r_1| + |r_2| + |r_3|)\,\epsilon, \qquad (|r_1 r_2| + |r_1 r_3| + |r_2 r_3|)\epsilon.$$

Apply this result to Exercise 1, where $\epsilon = \frac{1}{2}(10^{-4})$. How do the actual errors in

$$r_1 + r_2 + r_3, \qquad r_1 r_2 + r_1 r_3 + r_2 r_3, \qquad r_1 r_2 r_3,$$

compare with the maximum possible errors?

14. Two parallel walls are situated X feet apart; a ladder of length 30 feet stands with its foot at the base of one wall and leans against the second wall; another ladder of length 40 feet has its foot at the base of the second wall and leans against the first wall. If the point of intersection of the ladders is 10 feet above the ground, find X by showing

(a) $10\left[\dfrac{1}{\sqrt{900 - X^2}} + \dfrac{1}{\sqrt{1600 - X^2}}\right] = 1.$

Such an equation is very cumbersome, but can be simplified by using a scale factor. Put $X = 10z$ and show

(b) $\dfrac{1}{\sqrt{9 - z^2}} + \dfrac{1}{\sqrt{16 - z^2}} = 1.$

Clearly it would be troublesome to get rid of both surds by squaring; so put $v^2 = 9 - z^2$ and show that

(c) $v^4 - 2v^3 + 7v^2 - 14v + 7 = 0.$

This exercise shows that judicious manipulation of the initial result is very desirable before starting a numerical solution. Now solve the above equation by Newton's Method, noting that the desired root v is less than 3 (why?).

8. NEWTON'S METHOD WITH MULTIPLE OR NEAR-MULTIPLE ROOTS

It has sometimes been stated that Newton's method fails if α is a multiple root, since then $f(\alpha) = f'(\alpha) = 0$; we shall now show how to proceed in such a case.

EXAMPLE 4.15.

$$f(x) = x^7 + 6x^6 + 6x^5 - 21x^4 - 36x^3 + 3x^2 + 13x + 3 = 0,$$

that is,

$$f(x) = x + 6]x + 6]x - 21]x - 36]x + 3]x + 13]x + 3.$$

The graph of $f(x)$ is shown in Figure 4.7, and no difficulty is encountered in ascertaining its general features; $f(x)$ is always negative for $x < -3$, always positive for $x > 2$. In the intervening region, we find

x	$f(x)$
−3	−9
−2.5	−.0078125
−2	5
−1.5	11.1328125
−1	7
−.5	.3359375
0	3
.5	4.7265625
1	−25
1.5	−67.5703125
2	121

It is also important to note that $f(-2.6) = -.0025216$.

From Figure 4.7, we immediately find the following roots:

(a) a root α near .7;

(b) a root β near 1.9;

(c) possibly two roots γ and δ near −.4;

(d) a root ϵ near −2.5;

(e) possibly two roots ξ and η near −2.6.

The existence of the roots γ and δ is uncertain at the moment; the graph may just fail to touch the x-axis (γ and δ complex), it may touch the axis ($\gamma = \delta$), or it may cut the axis twice (γ nearly equal to δ). Likewise, the existence of ξ and η is by no means certain; the graph of $f(x)$ is suspiciously flat in the neighbourhood of ϵ. So ϵ may be a triple root, or the graph may have a tiny hump which just cuts the x-axis (ξ and η real) or just fails to cut the x-axis (ξ and η complex).

The single roots are easily found by Newton's method. Thus

$$\alpha = .65662, \qquad \beta = 1.83424, \qquad \epsilon = -2.49086.$$

The (possibly) double roots cause a little trouble. One method of attack is to note that if γ is a root of multiplicity n, then

(4.4) $$f(x) = (x - \gamma)^n A(x),$$

and

(4.5) $$f'(x) = (x - \gamma)^{n-1}[nA(x) + (x - \gamma)A'(x)].$$

It follows that γ is a root of the greatest common divisor of $f(x)$ and $f'(x)$. We can then dismiss the problem with the cavalier remark that there is a well-known method for finding the g.c.d. of two polynomials; unfortunately, this well-known method is not very practical for numerical work; in the present example, any attempt at finding the g.c.d. of $f(x)$ and $f'(x)$ soon bogs down.

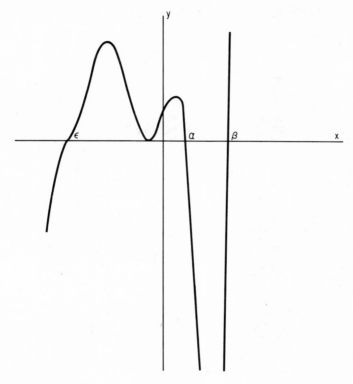

Fig. 4.7.

A feasible method of attack, however, becomes apparent if we use a localized approach; in the immediate vicinity of $x = \gamma$, (4.4) can be written

(4.6) $$f(x) = A(x - \gamma)^n,$$

where $A \doteq A(\gamma)$ is effectively constant. Then

(4.7) $$f'(x) = nA(x - \gamma)^{n-1},$$

$$f''(x) = n(n - 1)A(x - \gamma)^{n-2}, \qquad \text{etc.}$$

We thus find

(4.8) $$\frac{f'(x)}{f(x)} = \frac{n}{x - \gamma}$$

in the immediate neighbourhood of γ. Solving, we have the result

(4.9)
$$\gamma = x - n\frac{f(x)}{f'(x)},$$

where x is close to γ. This is the modification of Newton's rule for a multiple root, and we might state it as

THEOREM 4.1. If x_1 is in the neighbourhood of a root γ of multiplicity n, then

$$x_2 = x_1 - n\frac{f(x_1)}{f'(x_1)}$$

is an even closer approximation to γ.

Newton's method, as outlined in Section 7, is the case $n = 1$ of this theorem.

Using this theorem, we find a practical means for testing a multiple root in the

COROLLARY. If x_1 is in the neighbourhood of a root γ of the equation $f(x) = 0$, then γ will have multiplicity n if the expressions

$$x_1 - n\frac{f(x_1)}{f'(x_1)}, \quad x_1 - (n-1)\frac{f'(x_1)}{f''(x_1)}, \quad x_1 - (n-2)\frac{f''(x_1)}{f'''(x_1)}, \quad \ldots$$

are all effectively equal. (For these expressions are the approximations to an n-fold root of $f(x)$, an $(n-1)$-fold root of $f'(x)$, etc.)

Applying this theory in the present problem, take $x_1 = -.4$. Then

$$f'(x) = 7x + 36]x + 30]x - 84]x - 108]x + 6]x + 13,$$
$$f''(x) = 42x + 180]x + 120]x - 252]x - 216]x + 6.$$

We find

$$f(-.4) = .0078976, \quad f'(-.4) = -.875968,$$
$$f''(-.4) = 48.57792.$$

Then

$$x_1 - 2\frac{f(x_1)}{f'(x_1)} = -.38197, \quad x_1 - \frac{f'(x_1)}{f''(x_1)} = -.38197.$$

We conclude that $-.38197$ is (to five decimals) a double root of $f(x)$.

Similarly, we have

$$f(-2.6) = -.0025216, \quad f'(-2.6) = -.252928, \quad f''(-2.6) = -9.64992.$$

Here

$$x_1 - 2\frac{f(x_1)}{f'(x_1)} = -2.620, \quad x_1 - \frac{f'(x_1)}{f''(x_1)} = -2.626.$$

The possibility of a double root is still indicated; let us now take $x_2 = -2.62$. Then

$$f(-2.62) = .00003648833408, \quad f'(-2.62) = .037453642688,$$
$$f''(-2.62) = -19.5618442944.$$

Then

$$x_2 - 2\frac{f(x_2)}{f'(x_2)} = -2.61805, \quad x_2 - \frac{f'(x_2)}{f''(x_2)} = -2.61809.$$

A further iteration establishes -2.61803 as a double root.

EXAMPLE 4.16. If the roots are nearly equal, that is, the curve actually crosses the axis, the situation is somewhat simpler. Consider again the equation

$$f(x) = x - 27]x + 105]x - 140]x + 81]x - 21]x + 2 = 0$$

of Examples 2.4, 4.9, 4.14. There appears (Example 4.4) to be a near-double root near $x = 1/2$. In cases like this, it is well to tabulate the function; Table 4.5 gives functional values in $0(.1)1$, and the function is graphed in Figure 4.8.

Table 4.5

x	$10^6 f(x)$
0	2000000
.1	580231
.2	79424
.3	−4381
.4	15616
.5	−15625
.6	−124864
.7	−239741
.8	−217216
.9	128711
1.0	1000000

From Figure 4.8, we can immediately read off approximate roots as .27, .34, .47, and .88; using Newton's method, we find

$$x_1 = .27, \quad x_2 = .2667, \quad x_3 = .26695, \quad x_4 = .266931.$$
$$x_1 = .34, \quad x_2 = .328, \quad x_3 = .32760, \quad x_4 = .327597.$$
$$x_1 = .47, \quad x_2 = .476, \quad x_3 = .4757, \quad x_4 = .475693.$$
$$x_1 = .88, \quad x_2 = .8749, \quad x_3 = .874802.$$

Since we earlier found that 22.625853 and 2.429124 are roots, a check is provided by the fact that the sum of the six roots should be 27; this is indeed the case.

We should note two things in the computations; first, since $f'(x)$ is small, the convergence of Newton's method is slightly slower than usual, three or four iterations being needed to give six decimal places; secondly, since the roots are less than 1, the arrangement

$$f(x) = x - 27]x + 105]x - 140]x + 81]x - 21]x + 2$$

multiplies the round-off error at any stage by x, that is, keeps it less than 1 unit in the last place retained.

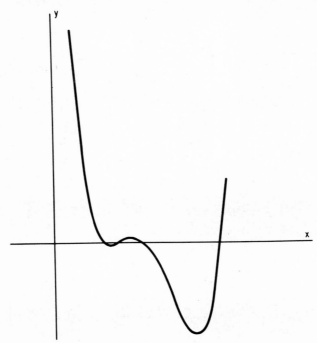

Fig. 4.8.

In concluding this section, a word is needed concerning the case when the two roots are very close, say at $\sigma \pm \epsilon$, where ϵ is very minute in relation to σ. The best attack here is to determine σ as a root of $f'(x) = 0$, and then apply Newton's method twice, once just to the right of σ and once just to the left of σ.

EXERCISES

1. For Example 4.15, write $f'(x)$, $f''(x)$, and $f'''(x)$ in bracket form. Solve the equations

(a) $f'(x) = 0$,   (b) $f''(x) = 0$,   (c) $f''(x) = 0$.

(Note that the graph of $f(x) = 0$ gives us approximate locations for the roots of $f'(x) = 0$.)

2. Graph the function

$$x^8 - 11x^7 + 38x^6 - 25x^5 - 88x^4 + 127x^3 - 62x^2 + 13x - 1 = 0,$$

and find the two pairs of triple roots.

3. Show that if α is a triple root of $f(x) = 0$, and if a and b are values close to α with $a < \alpha < b$, then α can be found by the *regula falsi* (this is convenient for any root of multiplicity n, n odd).

4. Solve completely the equation

$$x^6 + 12x^5 + 6x^4 - 208x^3 + 12x^2 + 48x + 8 = 0.$$

5. Let

$$f(x) = x^6 + 8x^5 + 39x^4 + 117x^3 + 206x^2 + 223x - 90;$$

use Theorem 4.1 to obtain the greatest common divisor of $f(x)$ and $f'(x)$. Hint: This will be $(x - \alpha)(x - \beta)$, where α and β are the double roots of $f(x) = 0$.

9. COMPLEX ROOTS

The case of an equation with complex coefficients is never easy (Newton's method is probably best); however, if the coefficients are real, we can gainfully combine a knowledge of the properties of the roots with Newton's method.

EXAMPLE 4.17. Solve the equation

$$x^4 - 2x^3 - 4x^2 + 5x - 7 = 0,$$

giving the roots to four decimals. (This equation is solved in Eshbach, *Handbook of Engineering Fundamentals*, second edition, p. 2–18, by the Graeffe process; comparison indicates that the Graeffe method should be avoided.) We at once compute

x	$f(x)$
-2	-1
-1	-13
0	-7
1	-7
2	-13
3	-1

and find that there are two real and two complex roots (Figure 4.9). We have

$$f(x) = \overline{x - 2]x - 4]x + 5]x} - 7,$$
$$f'(x) = \overline{4x - 6]x - 8]x} + 5.$$

Then

$$x_1 = 3, \qquad f(x_1) = -1, \qquad f'(x_1) = 35.$$
$$x_2 = 3.03, \quad f(x_2) = .07907081, \quad f'(x_2) = 36.947108.$$
$$x_3 = 3.03 - .0021 = 3.0279.$$

Since the graph is symmetrical, -2.0279 is also a root. Let the other roots

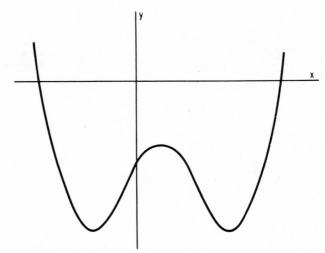

Fig. 4.9.

be $a \pm ib$. Since the sum of the roots is 2, we find $2a + 1 = 2$, $a = .5$. Also, the product of the roots is -7, that is,

$$(a^2 + b^2)(-6.14028) = -7, \quad b^2 = 1.1400 - .25 = .8900.$$

The two complex roots are thus $.5 \pm .9434i$. (See also Exercise 1.)

EXAMPLE 4.18.
$$x^4 - 5x^3 + 20x^2 - 40x + 60 = 0.$$

A graph of the function shows that there are no real roots; hence there must be two pairs of complex roots, $a \pm bi$ and $c \pm di$. Make a grid of functional values for points with integral coordinates in the complex plane (Figure 4.10). From this table of functional values, computed from

$$f(x) = x - 5]x + 20]x - 40]x + 60,$$

we see there is a root near $2 + 2i$. Using

$$f'(x) = 4x - 15]x + 40]x - 40,$$

we find, by Newton's method, which also holds for complex roots,

$$x_1 = 2 + 2i, \qquad f(x_1) = -4, \quad f'(x_1) = -24 + 24i.$$
$$x_2 = 1.92 + 1.92i, \qquad f(x_2) = -.12288i - .37929984,$$
$$f'(x_2) = -19.823104 + 22.831104i.$$
$$x_3 = 1.915 + 1.908i.$$

Two roots are then $1.915 \pm 1.908i$. If the other roots are $c \pm di$, we have

$$3.830 + (c + di) + (c - di) = 5,$$
$$(c^2 + d^2)(1.915^2 + 1.908^2) = 60.$$

These equations give the other two roots as $.585 \pm 2.805i$. If the four roots

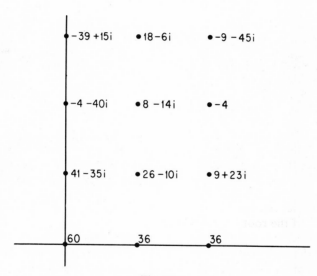

Fig. 4.10.

are desired more accurately, one further application of Newton's method should suffice.

EXERCISES

1. In Example 4.17, it is clear that $x = 1/2$ is an axis of symmetry of the curve; put $z = x - \frac{1}{2}$, and transform the equation to the form

$$z^4 - 5.5z^2 - 5.6875 = 0.$$

Deduce that

$$z^2 = 2.75 \pm \tfrac{1}{2}\sqrt{53} = 6.390055 \quad \text{or} \quad -.890055;$$

thence

$$z = \pm 2.527856 \quad \text{or} \quad \pm .94343i.$$

2. Solve the equations

    (a) $x^4 - 4x^3 + 10x^2 - 2x + 80 = 0,$
    (b) $x^3 - x^2 + 12x - 38 = 0,$
    (c) $x^4 + 2x^3 + 12x^2 - 2x - 10 = 0,$
    (d) $x^4 + 2x^3 + 8x^2 + x + 100 = 0.$

3. Show that the quartic equation

$$f(x) = x^4 + ax^3 + bx^2 + cx + d = 0$$

can have at most two real roots if $3a^2 - 8b < 0$. Hint: consider $f''(x)$.

4. Solve the equation

$$x^4 + 7x^3 + 30x^2 + 90x + 50 = 0$$

by (a) finding the two real roots α and β, and then using the properties of the roots of the equation, as in Example 4.17;

    (b) dividing out the factors $(x - \alpha)$ and $(x - \beta)$, and solving the real quadratic which results.

5. For the equation

$$x^4 - 25x^3 + 106x^2 - 200x + 115 = 0,$$

    (a) use an iterative procedure to determine the largest root to four decimals;
    (b) use Newton's Method to determine the other real root to four decimals;
    (c) use the properties of the roots to find the two complex roots.

10. AN ITERATIVE PROCEDURE FOR COMPLEX ROOTS

Suppose that $g(x) = 0$ is a polynomial equation with real coefficients and at least one pair of complex roots. If $x - \alpha$ (α real) is a factor of $g(x)$, then α can be determined and the factor $x - \alpha$ removed from $g(x)$ by division. Assuming that this has been done for all real roots, we see that there is no loss in generality in taking the initial equation in the form

(4.10) $f(x) = a_0x^n + a_1x^{n-1} + \dots + a_{n-1}x + a_n = 0,$

where $f(x) = 0$ has no real roots. However, since complex roots must occur in conjugate pairs of the form $\beta \pm i\theta$, we see that to each such pair of conjugate complex roots there corresponds a real quadratic factor of $f(x)$, namely, $x^2 - 2\beta x + (\beta^2 + \theta^2)$.

Our procedure in this section will be to determine a real quadratic factor $x^2 + ux + v$ of the polynomial $f(x)$ by a method of successive approximations. Suppose that we start with an arbitrary guess at the values of u and v. Then we can write

(4.11) $f(x) = (x^2 + ux + v)(b_0x^{n-2} + b_1x^{n-3} + \dots + b_{n-2}) + b_{n-1}x + B_n.$

(The constant term in the remainder is designated by a capital letter, since we shall find that it is given by a different formula from that used for the other b's.) Multiplying out the right-hand side of (4.11), and comparing coefficients, we obtain

(4.12)
$$\begin{cases} b_0 = a_0, \quad b_1 = a_1 - ub_0, \quad b_2 = a_2 - ub_1 - vb_0, \quad \dots, \\ b_{n-2} = a_{n-2} - ub_{n-3} - vb_{n-4}, \quad b_{n-1} = a_{n-1} - ub_{n-2} - vb_{n-3}, \\ \qquad\qquad B_n = a_n - vb_{n-2}. \end{cases}$$

Formulae (4.12) can all be resumed in the single formula

(4.13) $b_k = a_k - ub_{k-1} - vb_{k-2} \quad (k = 0, 1, \dots, n), \quad B_n = a_n - vb_{n-2},$

if we adopt the obvious formal convention that $b_{-1} = b_{-2} = 0$. (We shall find it convenient to use b_n later, although it has not yet appeared.)

Now it is clear that algorithm (4.12) provides us with a quadratic factor $x^2 + ux + v$ if and only if the coefficients b_{n-1} and B_n in the remainder vanish. It is highly improbable that the first trial values of u and v will make b_{n-1} and B_n vanish; so we attempt to alter u to a new value $u + du$ and v to a new value $v + dv$ in such a way that the new values of b_{n-1} and B_n will be nearly zero. These new values of b_{n-1} and B_n will be approximately

(4.14)
$$b_{n-1} + db_{n-1} = b_{n-1} + \frac{\partial b_{n-1}}{\partial u} du + \frac{\partial b_{n-1}}{\partial v} dv$$

and

(4.15)
$$B_n + dB_n = B_n + \frac{\partial B_n}{\partial u} du + \frac{\partial B_n}{\partial v} dv$$

respectively, and these new approximate values should be zero if $u + du$ and $v + dv$ are improved coefficients for our quadratic factor. Consequently, to use Formulae (4.14) and (4.15), we must determine the partial derivatives of b_{n-1} and B_n with respect to u and v.

To obtain the four partial derivatives needed, write (4.11) in the form

(4.16)
$$f(x) = (x^2 + ux + v)q(x) + b_{n-1}x + B_n,$$

and differentiate with respect to u and v. Then

(4.17)
$$0 = (x^2 + ux + v)\frac{\partial q(x)}{\partial u} + xq(x) + x\frac{\partial b_{n-1}}{\partial u} + \frac{\partial B_n}{\partial u},$$

and

(4.18)
$$0 = (x^2 + ux + v)\frac{\partial q(x)}{\partial v} + q(x) + x\frac{\partial b_{n-1}}{\partial v} + \frac{\partial B_n}{\partial v}.$$

Writing Equations (4.17) and (4.18) in the form

(4.19)
$$\begin{cases} xq(x) = -(x^2 + ux + v)\dfrac{\partial q(x)}{\partial u} - x\dfrac{\partial b_{n-1}}{\partial u} - \dfrac{\partial B_n}{\partial u}, \\[3mm] q(x) = -(x^2 + ux + v)\dfrac{\partial q(x)}{\partial v} - x\dfrac{\partial b_{n-1}}{\partial v} - \dfrac{\partial B_n}{\partial v}, \end{cases}$$

we see that the remainders when $xq(x)$ and $q(x)$ are divided by $x^2 + ux + v$ are given by

(4.20)
$$-x\frac{\partial b_{n-1}}{\partial u} - \frac{\partial B_n}{\partial u}, \quad -x\frac{\partial b_{n-1}}{\partial v} - \frac{\partial B_n}{\partial v},$$

respectively. Consequently, to obtain these quantities in (4.20), we write

(4.21)
$$q(x) = b_0 x^{n-2} + \dots + b_{n-3}x + b_{n-2}$$
$$= (x^2 + ux + v)(c_0 x^{n-4} + \dots + c_{n-5}x + c_{n-4}) + c_{n-3}x + C_{n-2}.$$

Comparing coefficients in (4.21) gives a completely analogous result to (4.13), namely,

(4.22)
$$c_k = b_k - uc_{k-1} - vc_{k-2} \quad (k = 0, 1, \dots, n-3), \quad C_{n-2} = b_{n-2} - vc_{n-4}.$$

Also it is immediately verified that

(4.23)
$$xq(x) = (x^2 + ux + v)(c_0 x^{n-3} + \dots + c_{n-4}x + c_{n-3}) + c_{n-2}x - vc_{n-3},$$

where we extend (4.22) by defining

$$c_{n-2} = b_{n-2} - uc_{n-3} - vc_{n-4} = C_{n-2} - uc_{n-3}.$$

Equating the two forms of remainder found in (4.20) and in (4.21), (4.23), we have the identities

(4.24)
$$\begin{cases} -x\dfrac{\partial b_{n-1}}{\partial u} - \dfrac{\partial B_n}{\partial u} = c_{n-2}x - vc_{n-3}, \\[3mm] -x\dfrac{\partial b_{n-1}}{\partial v} - \dfrac{\partial B_n}{\partial v} = c_{n-3}x + c_{n-2} + uc_{n-3}. \end{cases}$$

Hence

(4.25)
$$\begin{cases} \dfrac{\partial b_{n-1}}{\partial u} = -c_{n-2}, \quad \dfrac{\partial B_n}{\partial u} = vc_{n-3}, \\[3mm] \dfrac{\partial b_{n-1}}{\partial v} = -c_{n-3}, \quad \dfrac{\partial B_n}{\partial v} = -c_{n-2} - uc_{n-3}. \end{cases}$$

Substituting from (4.25) in (4.14) and (4.15), and equating these coefficients to zero, we obtain

(4.26)
$$\begin{cases} c_{n-2}\,du + c_{n-3}\,dv = b_{n-1}, \\ vc_{n-3}\,du - (c_{n-2} + uc_{n-3})\,dv = -B_n = -b_n - ub_{n-1}. \end{cases}$$

Equations (4.26) are readily solved to give

(4.27)
$$du = \frac{1}{\Delta}\begin{vmatrix} b_{n-1} & c_{n-3} \\ b_n & c_{n-2} \end{vmatrix}, \qquad dv = \frac{-1}{\Delta}\begin{vmatrix} b_{n-1} & c_{n-2} \\ b_n & c_{n-1} \end{vmatrix},$$

where Δ is the determinant

(4.28)
$$\Delta = \begin{vmatrix} c_{n-2} & c_{n-3} \\ c_{n-1} & c_{n-2} \end{vmatrix},$$

and we have set

$$-uc_{n-2} - vc_{n-3} = c_{n-1}.$$

In summary, our procedure should then be as follows: write out

$$a_k\,(k = 0, 1, \ldots, n), \quad b_k\,(k = 0, 1, \ldots, n), \quad c_k\,(k = 0, 1, \ldots, n-1),$$

in three parallel vertical columns. Compute du and dv from (4.27). Then $u + du, v + dv$, is an improved pair of coefficients for the desired quadratic factor. Repeat this procedure with $x^2 + (u + du)x + (v + dv)$ as trial divisor.

It is clear from Formulae (4.27) and (4.28) that du and dv will tend to approach the indeterminate form 0/0 in case $(x^2 + ux + v)$ is a repeated factor of $f(x)$; in that case $(x^2 + ux + v)$ is also a factor of $q(x)$ and the coefficients c_{n-2} and c_{n-3}, which determine the remainder in (4.21), will approach zero as b_{n-1} and b_n approach zero. In this case, $x^2 + ux + v$ can be determined as a factor of $f'(x)$.

EXAMPLE 4.19. Use the method of this section on Example 4.18,
$$x^4 - 5x^3 + 20x^2 - 40x + 60 = 0.$$

Since we know nothing about either quadratic factor, let us take $u = v = 0$ as our initial values. Also, it will be convenient to write (4.27) in the form $du = \Delta_1/\Delta,\ dv = -\Delta_2/\Delta.$

n	a	b	c
0	1	1	1
1	-5	-5	-5
2	20	20	20
3	-40	-40	0
4	60	60	

$$\Delta_1 = \begin{vmatrix} -40 & -5 \\ 60 & 20 \end{vmatrix} = -500;$$
$$\Delta_2 = \begin{vmatrix} -40 & 20 \\ 60 & 0 \end{vmatrix} = 1200;$$
$$\Delta = \begin{vmatrix} 20 & -5 \\ 0 & 20 \end{vmatrix} = 400;$$
$$du = -1.25; \quad dv = +3.$$

Our second trial divisor is thus $x^2 - 1.25x + 3$. Repeat the procedure.

n	a	b	c
0	1	1	1
1	−5	−3.75	−2.5
2	20	12.3125	6.1875
3	−40	−13.359375	15.234375
4	60	6.363281	

$\Delta = 76.371;$ $\Delta_1 = -66.753;$
$\Delta_2 = -242.895;$ $du = -.87;$ $dv = 3.18.$

Third trial divisor: $x^2 - 2.12x + 6.18$.

a	b	c
1	1	1
−5	−2.88	−.76
20	7.7144	−.0768
−40	−5.847072	4.533984
60	−.070785	

$\Delta = 3.452;$ $\Delta_1 = .395;$
$\Delta_2 = -26.516;$ $du = .11;$ $dv = 7.68.$

Fourth trial divisor: $x^2 - 2.01x + 13.86$.

a	b	c
1	1	1
−5	−2.99	−.98
20	.1301	−15.6997
−40	1.702901	−17.973597
60	61.619645	

$\Delta = 228.866;$ $\Delta_1 = 33.652;$
$\Delta_2 = 936.803;$ $du = .15;$ $dv = -4.09.$

Fifth trial divisor: $x^2 - 1.86x + 9.77$.

a	b	c
1	1	1
−5	−3.14	−1.28
20	4.3896	−7.7612
−40	−1.157544	−1.930232
60	14.960576	

$\Delta = 57.766;$ $\Delta_1 = 28.133;$
$\Delta_2 = 118.346;$ $du = .49;$ $dv = -2.05.$

Sixth trial divisor: $x^2 - 1.37x + 7.72$.

a	b	c
1	1	1
−5	−3.63	−2.26
20	7.3069	−3.5093
−40	−1.965947	12.639459
60	.897385	

$\Delta = 40.880;$ $\Delta_1 = 8.927;$
$\Delta_2 = -21.699;$ $du = .22;$ $dv = .53.$

Seventh trial divisor: $x^2 - 1.15x + 8.25$.

a	b	c
1	1	1
−5	−3.85	−2.70
20	7.3225	−4.0325
−40	.183375	17.637625
60	−.199744	

$\Delta = 63.8826;$ $\Delta_1 = -1.2788;$
$\Delta_2 = 2.4288;$ $du = -.0200;$
$dv = -.0380.$

Eighth trial divisor: $x^2 - 1.1700x + 8.2120$.

a	b	c
1	1	1
−5	−3.83	−2.66
20	7.3069	−4.0173
−40	.001033	17.143679
60	−.003054	

$\Delta = 61.740885;$ $\Delta_1 = -.012274;$
$\Delta_2 = .005441;$ $du = -.000199;$
$dv = -.000088.$

Ninth trial divisor: $x^2 - 1.170199x + 8.211912$.

a	b	c
1	1	1
−5	−3.829801	−2.659602
20	7.306459	−4.017707
−40	.000000	17.138901
60	.000002	

$\Delta = 61.724625;$ $\Delta_1 = .000005;$
$\Delta_2 = .000008;$ $du = dv = .000000.$

We have finally reached a stage where the values of Δ_1 and Δ_2 are effectively zero; so we can read the two factors of $f(x)$ from the last tableau as

$$x^2 - 1.170199x + 8.211912 \quad \text{and} \quad x^2 - 3.829801x + 7.306459.$$

The corresponding roots are

$$.5850995 \pm 2.8052755i \quad \text{and} \quad 1.9149005 \pm 1.9077775i.$$

Example 4.19 illustrates two important points; first, a minor error at any stage is self-correcting, since we always divide our trial divisor into the original polynomial; secondly, convergence is much faster if we start out close to the required factor. To stress this second point, let us consider

EXAMPLE 4.20. Take the same function

$$x^4 - 5x^3 + 20x^2 - 40x + 60$$

as in Example 4.19. Suppose, however, that from a grid or some other considerations, it is known that $2 + 2i$ is an approximate root. Then an approximate factor is $x^2 - 4x + 8$; using it as our first trial divisor, we have

a	b	c
1	1	1
−5	−1	3
20	8	12
−40	0	24
60	−4	

$\Delta = 72; \quad \Delta_1 = 12; \quad \Delta_2 = 48; \quad du = .17; \quad dv = -.67.$

From the second trial divisor, $x^2 - 3.83x + 7.33$, we obtain

a	b	c
1	1	1
−5	−1.17	2.66
20	8.1889	11.0467
−40	−.060413	22.811061
60	−.256019	

$\Delta = 61.352; \quad \Delta_1 = .01365;$
$\Delta_2 = 1.4501; \quad du = .0002; \quad dv = -.0236.$

Two iterations have thus sufficed to obtain $x^2 - 3.8298x + 7.3064$ as an approximate factor.

EXERCISES

1. Solve completely the equation

$$x^5 + 9x^3 - 8x^2 + 20 = 0.$$

2. Solve completely the equation

$$x^6 + 2x^5 + 10x^4 + 10x^3 - 4x^2 + 6x + 60 = 0.$$

Hint: It is best to locate an approximate root if possible, and then improve the value as in Example 4.20.

3. Solve completely the equation

$$x^6 - x^5 - 3x^4 - 2x^3 - 2x^2 - 12x + 12 = 0.$$

4. Suppose, in analogy to the development of this section, that

$$f(x) = a_0 x^n + a_1 x^{n-1} + \ldots + a_{n-1} x + a_n = (x + u)q(x) + b_n,$$

where

$$q(x) = b_0 x^{n-1} + \ldots + b_{n-2} x + b_{n-1}.$$

Suppose further that

$$q(x) = (x + u)(c_0 x^{n-2} + \ldots + c_{n-3} x + c_{n-2}) + c_{n-1}.$$

Prove the following results.

(a) $b_k = a_k - ub_{k-1} \quad (k = 0, 1, \ldots, n)$.
(b) $c_k = b_k - uc_{k-1} \quad (k = 0, 1, \ldots, n - 1)$.
(c) When $q(x)$ is divided by $x + u$, the remainder is $-db_n/du$.
(d) $du = b_n/c_{n-1}$.

5. Use the method of Exercise 4, starting from $u = 0$, to obtain a root of the equation
$$x^4 + x^3 + x^2 + 10x - 10 = 0.$$

6. Use the method of Exercise 4 to improve the approximate root .66 of the equation
$$x^3 - 5x + 3 = 0.$$

7. Using the methods of this section and Exercise 4, factor completely the polynomial
$$x^7 + 2x^2 + 4x - 20.$$

8. Solve the equation
$$x^4 + 4x^3 + 9x^2 + x + 10 = 0$$

by the following method.

(a) Eliminate x^3 by the substitution $x = y - 1$ to give
$$y^4 + 3y^2 - 9y + 15 = 0.$$

(b) Let $y = p + iq$ be a root; show that p^2 is the positive root of
$$64z^3 + 96z^2 - 204z - 81 = 0$$
and that
$$q^2 = p^2 + \frac{3}{2} - \frac{9}{4p}.$$

(c) Let $4z + 2 = K$; obtain K as the largest root of
$$K^3 - 63K + 37 = 0.$$

Thence deduce the roots of the original equation. (Use Newton's method with $K = 8$.)

(d) Check your result by writing down the two quadratic factors of
$$x^4 + 4x^3 + 9x^2 + x + 10,$$

and then multiplying them.

CHAPTER FIVE

Computation with Series
and Integrals

1. REVIEW OF SERIES

Let a series

(5.1)
$$u_1 + u_2 + u_3 + \dots + u_n + \dots$$

be given, in such a way that the law of formation of the nth term is known. Let

$$S_n = \sum_{j=1}^{n} u_j$$

represent the sum of n terms. If

(5.2)
$$\lim_{n \to \infty} S_n$$

does not exist, then the series behaves indefinitely, and we do not (at this stage of our work) consider it further. (Actually, there are numerous methods available whereby a value may be assigned to an indefinite series. See, for example, Konrad Knopp, *Theory and Application of Infinite Series.*) If the limit (5.2) does exist, it may be either finite or infinite. In either case, we say that the series behaves *definitely*; when the limit of S_n is infinite, we call the series *divergent*; when the limit of S_n is a finite number α, we say that the series is *convergent* to the sum α. We shall be concerned almost exclusively with convergent series.

For a detailed discussion of convergent series, one may refer to almost any elementary calculus text (see, for example, *Elements of Calculus*, by T. S. Peterson). Here we shall list a number of basic properties.

1) Any *finite* number of terms in a series may be altered, omitted, or rearranged without affecting convergence.

2) Parentheses may be inserted or removed without affecting the value of the series.

3) An infinite number of terms may be rearranged without altering the value of the series, *provided* that the series is absolutely convergent (that is, remains convergent when all terms are made positive).

4) A series of positive terms is convergent (divergent) if each term is smaller (larger) than the corresponding term of a convergent (divergent) series.

5) If the series (5.1) converges, then $u_n \to 0$ as $n \to \infty$.

6) A series converges if the Cauchy ratio

$$R = \operatorname*{Lim}_{n \to \infty} \left| \frac{u_{n+1}}{u_n} \right|$$

is less than 1; it diverges if $R > 1$.

7) The geometric series $\sum\limits_{n=0}^{\infty} ax^n$ converges to the value $a/(1 - x)$ if and only if $|x| < 1$.

EXERCISES

1. Find the sum of n terms of the series whose nth term is

   (a) $\log \dfrac{n}{n + 1}$,   (b) $\dfrac{1}{n(n + 3)}$,   (c) $\dfrac{1}{n(n + 1)(n + 2)}$,

   (d) $(-)^n$,   (e) $n^2 - 3n + 2$,   (f) 2^{-n}.

   Determine from S_n whether or not the series converges; if it does, find its sum.

2. It is known that the series

   $$S = 1 - \tfrac{1}{2} + \tfrac{1}{3} - \tfrac{1}{4} + \tfrac{1}{5} - + \ldots$$

   converges to $\log 2$; assuming this result, show that the series

   $$T = 1 + \tfrac{1}{3} - \tfrac{1}{2} + \tfrac{1}{5} + \tfrac{1}{7} - \tfrac{1}{4} + \tfrac{1}{9} + \tfrac{1}{11} - \tfrac{1}{6} + + - \ldots$$

   has the value $\tfrac{3}{2}S$. (Note that T is S with an infinite number of terms rearranged.)

3. Find the nth term of the series

   $$1 + \tfrac{1}{5} + \tfrac{1}{6} + \tfrac{1}{31} + \tfrac{1}{128} + \tfrac{1}{369} + \tfrac{1}{850} + \ldots,$$

   given that, for $n > 1$, the nth term has the form $1/F(x)$, where $F(x)$ is a quartic polynomial.

2. ERROR BOUNDS IN SERIES COMPUTATIONS

A rough bound on the value of a series is provided by

THEOREM 5.1 (Maclaurin). If $f(x)$ is a monotonely decreasing function such that $\lim_{x \to \infty} f(x) = 0$, then

$$\int_1^\infty f(x)\,dx < \sum_1^\infty f(x) < f(1) + \int_1^\infty f(x)\,dx.$$

Proof. Consider a sequence of rectangles of unit width whose upper left vertices lie on the graph of the function $f(x)$ (Figure 5.1). Since the area under the curve is $\int_1^\infty f(x)\,dx$ and the area of the rectangles is $\sum_1^\infty f(x)$, we obtain the left-hand inequality in the enunciation of the theorem.

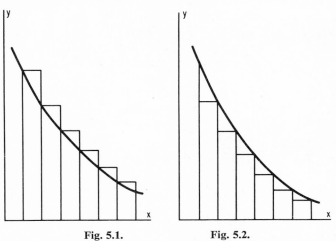

Fig. 5.1.   Fig. 5.2.

Now (Figure 5.2) consider a set of rectangles of unit width inscribed under the curve $f(x)$; in this case the area of the rectangles is $\sum_2^\infty f(x)$, and we obtain

$$\sum_2^\infty f(x) < \int_1^\infty f(x)\,dx.$$

Adding $f(1)$ to each side, we obtain the right-hand inequality in the theorem.

Quite clearly, this theorem is not capable of giving very exact results. For example, it merely tells us that $\sum_1^\infty x^{-2}$ lies between 1 and 2; the actual value is $\pi^2/6 = 1.64 \ldots$.

If the terms of a series satisfy a recursion relation, this can be employed to give a bound on the error. As an illustration, consider the rather hackneyed

EXAMPLE 5.1. Compute

$$e = \sum_{r=0}^{\infty} \frac{1}{r!}.$$

Clearly

$$e = u_0 + u_1 + u_2 + \dots \quad \text{where} \quad u_0 = 1, \quad u_{r+1} = \frac{u_r}{r+1}.$$

Then we obtain $u_0 = 1.00000$

$$
\begin{array}{ll}
u_1 = u_0/1 = 1.00000 & u_5 = u_4/5 = .00833^+ \\
u_2 = u_1/2 = .50000 & u_6 = u_5/6 = .00139^- \\
u_3 = u_2/3 = .16667^- & u_7 = u_6/7 = .00020^- \\
u_4 = u_3/4 = .04167^- & u_8 = u_7/8 = .00002^+
\end{array}
$$

Adding, $e \doteq 2.71828.$

How accurate is this value for e?

There are two sources of error: round-off error, occasioned by using only five figures in each term; and truncation error, occasioned by stopping the computations at u_8.

The round-off error comprises six round-offs, two in defect (marked $+$) and four in excess (marked $-$); since a round-off error is at most 1/2 unit in the last decimal place, the value obtained for e may lie between 2.71826 and 2.71829, as far as round-off is concerned.

The truncation error can be bounded above, since it is given by

$$u_9 + u_{10} + u_{11} + \dots = u_9 + \frac{u_9}{10} + \frac{u_9}{10 \cdot 11} + \frac{u_9}{10 \cdot 11 \cdot 12} + \dots$$

$$< u_9 + \frac{u_9}{10} + \frac{u_9}{10^2} + \frac{u_9}{10^3} + \dots = \frac{10}{9} u_9 = \frac{10}{81} u_8.$$

This is considerably less than 1/2 unit in the fifth decimal place. Truncation error thus is practically negligible as regards the first five decimals.

We conclude that $e = 2.7183$ to four decimals; the fifth decimal served as a guard to protect the fourth against the round-off error. (Occasionally, when more terms are to be added, two guard figures may be needed.)

The most useful type of series for computational purposes is the *alternating* series

$$S = u_1 - u_2 + u_3 - u_4 + u_5 - u_6 + \dots$$

where

$$u_n > 0, \quad u_n > u_{n+1}, \quad u_n \to 0 \quad \text{as} \quad n \to \infty.$$

It is clear that

$$S = (u_1 - u_2) + (u_3 - u_4) + \dots > 0,$$

and that

$$S = u_1 - (u_2 - u_3) - (u_4 - u_5) - \dots < u_1.$$

Hence

THEOREM 5.2. The value of an alternating series is less than the first term.

Since the truncation error in an alternating series is itself an alternating series, we have the

COROLLARY. The truncation error in an alternating series is less than the first term neglected.

EXAMPLE 5.2.

$$\sin x = x - \frac{x^3}{3!} + \frac{x^5}{5!} - \frac{x^7}{7!} + \cdots .$$

Let $x = .2$; then

$$u_1 = .2, \quad u_2 = -\frac{.04}{6} u_1 = -.00133, \quad u_3 = -\frac{.04}{20} u_2 = +.000003.$$

Using two terms of the series only,

$$\sin .2 = .2 - .00133 = .19867,$$

and the truncation error is less than 3×10^{-6}, that is, it does not affect the fifth decimal.

EXERCISES

1. Given

$$u(x) = \sum_{j=0}^{\infty} \frac{x^j}{(j + 1)(j + 2)(j + 3)},$$

express the nth term of the series in terms of the $(n - 1)$th term; thence compute $u(1/2)$, using five terms of the series. Give a bound for the truncation error made in stopping the series at the term in x^t; what is this error for $t = 4$?

2. Let

$$a_r = \frac{1^2 \, 3^2 \, 5^2 \cdots (2r - 1)^2}{4^r (2r)!} .$$

Find a recursion formula for a_r, and prove that the truncation error occasioned in replacing $\sum_{r=0}^{\infty} a_r$ by $\sum_{r=0}^{n} a_r$ is less than a_n. Use this result to compute $\sum_{r=0}^{\infty} a_r$ correct to three decimals. Check the numerical value obtained by verifying that

$$\sum_{0}^{\infty} a_r = \sum_{0}^{\infty} \binom{2r}{r} 2^{-4r} = \left(1 - \frac{1}{4}\right)^{-1/2} = \frac{2}{\sqrt{3}} .$$

(These coefficients a_r appear in Formula 3.17.)

3. Follow the method of Exercise 2 in discussing the numerical value of

$$\sum_{r=0}^{\infty} 2^r a_r.$$

4. (a) By comparing the area under the curve $y = 1/x$ from n to $n + 1$ with the area of the trapezoid formed by the tangent at $n + \frac{1}{2}$ and the ordinates at n and $n + 1$, prove

$$\frac{1}{n + \frac{1}{2}} < \log \frac{n + 1}{n}.$$

Hint: the trapezoid is equal in area to a rectangle formed by a horizontal line through $[n + \frac{1}{2}, (n + \frac{1}{2})^{-1}]$.

(b) Using (a), deduce that the sequence $\sum_{x=1}^{n} \frac{1}{x} - \log n$ is monotone decreasing.

(c) By considering the area under the curve $y = 1/x$, prove that

$$0 < \sum_{x=1}^{n} \frac{1}{x} - \log n < 1 \quad \text{for all } n.$$

Deduce that

$$\gamma = \lim_{n \to \infty} \left[\sum_{x=1}^{n} \frac{1}{x} - \log n \right]$$

exists.

5. Use the Maclaurin inequality to show that $\sum_{x=1}^{\infty} \frac{1}{x^p}$ converges for $p > 1$, and diverges if $p \leq 1$. In the case of convergence, show that the value of the series is less than $p/(p - 1)$.

6. By approximating the area under $y = (1 + x^2)^{-1}$ in the interval $(0, 1)$, prove that

$$\lim_{n \to \infty} n \sum_{x=0}^{n} \frac{1}{x^2 + n^2} = \frac{\pi}{4}.$$

7. (a) Let

$$k_n = \log n! - (n + \frac{1}{2}) \log n + n.$$

Use 4(a) to prove $k_{n+1} - k_n < 0$, that is, the sequence k_n is monotone decreasing.

(b) Compare the area under the curve $y = \log x$ with the area made up of
   (i) the triangle formed by the tangent to the curve at $(1, 0)$ and the line $x = 3/2$;
   (ii) the trapezoid formed by the tangent at r and the lines $x = r \pm \frac{1}{2}$ $(r = 2, \dots, n - 1)$; this is equal to a rectangle with top passing through $(r, \log r)$;
   (iii) a rectangle of height $\log n$ from $n - \frac{1}{2}$ to n.
   Deduce that $k_n > \frac{7}{8}$ and hence that $\lim_{n \to \infty} k_n$ exists.

8. How many terms of the series

$$\cos x = 1 - \frac{x^2}{2!} + \frac{x^4}{4!} - \frac{x^6}{6!} + \dots$$

must be taken to ensure that $\cos .5$ be computed correct to six decimals? Evaluate $\cos .5$ to six decimals by performing the computation in the form

$$\cos x = \dots - \frac{1}{720}]x^2 + \frac{1}{24}]x^2 - \frac{1}{2}]x^2 + 1.$$

3. REVIEW OF TAYLOR SERIES

Let $f(x)$ be any function, and suppose that it can be represented as a power series in $x - c$,

$$f(x) = a_0 + a_1(x - c) + a_2(x - c)^2 + a_3(x - c)^3 + \cdots$$

This is actually possible for almost all ordinary functions; there are some few pathological cases, which the student is not likely to meet. Assuming that the power series can be differentiated term by term, we find

$$f'(x) = 1!a_1 + 2a_2(x - c) + 3a_3(x - c)^2 + 4a_4(x - c)^3 + \cdots$$

$$f''(x) = 2!a_2 + 3 \cdot 2a_3(x - c) + 4 \cdot 3a_4(x - c)^2 + \cdots$$

$$f'''(x) = 3!a_3 + 4 \cdot 3 \cdot 2a_4(x - c) + \cdots$$

$$f''''(x) = 4!a_4 + \cdots$$

Putting $x = c$ in these relations gives the formula

$$f^{(n)}(c) = n!a_n;$$

hence

(5.3) $$f(x) = \sum_{n=0}^{\infty} \frac{f^{(n)}(c)}{n!}(x - c)^n.$$

If we put $x = c + h$, we get the useful alternative form

(5.4) $$f(c + h) = \sum_{n=0}^{\infty} \frac{f^{(n)}(c)}{n!} h^n.$$

This is Taylor's Series. For $c = 0$, we obtain

(5.5) $$f(x) = \sum_{n=0}^{\infty} \frac{f^{(n)}(0)}{n!} x^n.$$

This is known as Maclaurin's Series.

The Taylor series (5.3) will converge for a range of values

$$-r < x - c < +r;$$

r is called the radius of convergence. The series will diverge for $|x - c| > r$, and its behaviour for $|x - c| = r$ must be determined in each individual case. We recall two theorems.

1) The series (5.3) may be differentiated or integrated *within* its interval of convergence, and the result will be valid *within* that interval.

2) Two series may be added, subtracted, or multiplied, and the result will be valid *within* their common interval of convergence.

Applying (5.5) directly to some common functions, we obtain:

(5.6)
$$\sin x = \sum_{r=0}^{\infty} (-)^r \frac{x^{2r+1}}{(2r+1)!} \qquad \text{(all } x\text{)},$$

(5.7)
$$\cos x = \sum_{r=0}^{\infty} (-)^r \frac{x^{2r}}{2r!} \qquad \text{(all } x\text{)},$$

(5.8)
$$(1+x)^n = \sum_{r=0}^{\infty} \binom{n}{r} x^r \qquad (-1 < x < 1),$$

(5.9)
$$e^x = \sum_{r=0}^{\infty} \frac{x^r}{r!} \qquad \text{(all } x\text{)},$$

(5.10)
$$\log(1+x) = \sum_{r=1}^{\infty} (-)^{r+1} \frac{x^r}{r} \qquad (-1 < x \le 1).$$

The range of validity of these, or of any Taylor series, may always be determined from the Cauchy ratio R.

EXAMPLE 5.3. Determine the region of convergence of (5.10).

Since

$$|u_{r+1}| = \frac{x^{r+1}}{r+1} \quad \text{and} \quad |u_r| = \frac{x^r}{r},$$

we at once have

$$R = \operatorname*{Lim}_{r \to \infty} \left| \frac{rx}{r+1} \right| = |x|.$$

Hence the series (5.10) converges if $|x| < 1$, diverges if $|x| > 1$. For $|x| = 1$, we obtain $1 + \frac{1}{2} + \frac{1}{3} + \ldots$, which is the (divergent) harmonic series; or $1 - \frac{1}{2} + \frac{1}{3} - \ldots$, which is a (convergent) alternating series, and hence represents log 2. Consequently, the region of convergence is $-1 < x \le 1$.

The only difficulty with any of these formulae arises when we wish to know if (5.8) holds for $x = \pm 1$; this is a complicated question (see, for example, I. S. Sokolnikoff, *Advanced Calculus*).

One of the most important uses of power series occurs in evaluating integrals.

EXAMPLE 5.4. Evaluate $\int_0^{1/2} e^{-t^2/2} \, dt$, which occurs in probability.

Using (5.9), we have

$$\int_0^{1/2} e^{-t^2/2}\, dt = \int_0^{1/2}\left[1 - \frac{t^2}{2} + \frac{t^4}{2^2 2!} - \frac{t^6}{2^3 3!} + \cdots\right] dt$$

$$= \left[t - \frac{t^3}{6} + \frac{t^5}{40} - \frac{t^7}{336} + \cdots\right]_0^{1/2}$$

$$= \frac{1}{2} - \frac{1}{48} + \frac{1}{1280} - \frac{1}{43008} + \cdots$$

$$= .5 - .02083 + .00078 - .00002 + \cdots$$

$$= .47993,$$

with the truncation error $< \dfrac{1}{9 \cdot 2^{13} \cdot 4!} = .0000005.$

EXERCISES

1. Show that, for all x,

$$\sinh x = \sum_{r=0}^{\infty} \frac{x^{2r+1}}{(2r+1)!}, \qquad \cosh x = \sum_{r=0}^{\infty} \frac{x^{2r}}{2r!}.$$

Obtain these results directly from (5.5), and also by employing (5.9).

2. Obtain, to six places of decimals, the values of

(a) $\displaystyle\int_0^{.3} \sin x^2\, dx,$ (b) $\displaystyle\int_0^{.1} \log(1 + x^2)\, dx,$

(c) $\displaystyle\int_0^{.2} (1 + x^2)^{1/3}\, dx,$ (d) $\displaystyle\int_0^{.2} \tan x^3\, dx.$

3. (a) Obtain (5.10) by integrating (5.8) with $n = -1$, that is,

$$\int_0^X \frac{1}{1+x}\, dx = \int_0^X [1 - x + x^2 - x^3 + \cdots]\, dx.$$

(b) Obtain the series for $\tan^{-1} X$ from

$$\int_0^X \frac{1}{1+x^2}\, dx = \int_0^X [1 - x^2 + x^4 - x^6 + \cdots]\, dx.$$

Deduce the value of $\displaystyle\int_0^{1/5} \tan^{-1} x\, dx.$

4. The expression $\displaystyle\int_0^{\pi/2} \frac{dx}{\sqrt{1 - k^2 \sin^2 x}}$ occurs in discussing the motion of a pendulum ($k^2 < 1$).

(a) Use integration by parts to obtain a formula for

$$\int_0^{\pi/2} \sin^{2m} x \, dx.$$

(b) Deduce that the original expression equals

$$\frac{\pi}{2}\left[1 + \left(\frac{1}{2}\right)^2 k^2 + \left(\frac{1 \cdot 3}{2 \cdot 4}\right)^2 k^4 + \left(\frac{1 \cdot 3 \cdot 5}{2 \cdot 4 \cdot 6}\right)^2 k^6 + \dots\right].$$

(c) Show that, if $k < .2$, the approximation $\frac{\pi}{2}\left[1 + \frac{k^2}{4}\right]$ gives a value of the integral with a truncation error $< .0004$.

5. The Bernoulli numbers $B_0, B_1, B_2, \dots$ are defined by the formal relation

$$\frac{x}{e^x - 1} = \sum_{r=0}^{\infty} B_r \frac{x^r}{r!}.$$

Find, by cross-multiplying and equating coefficients, the values of $B_0, B_1, \dots, B_6$.

6. Show that the function

$$g(x) = \frac{x}{2} + \frac{x}{e^x - 1}$$

has the property $g(x) = g(-x)$, that is, $g(x)$ is an even function; deduce that $B_3 = B_5 = \dots = 0$.

7. Use the method employed in the proof of Theorem 3.1 to show that (5.4) can be written in the form

$$f(c + h) = \sum_{m=0}^{n} \frac{f^{(m)}(c)}{m!} h^m + \frac{f^{(n+1)}(c + \theta h)}{(n+1)!} h^{n+1},$$

where $0 \le \theta \le 1$. This form of (5.4) is known as Taylor's Series with Remainder; it shows that the error made in truncating the Taylor Series at any stage is less than the next term (altered by the evaluation of the requisite derivative, not at c, but at some place intermediate between c and $c + h$).

8. Use the result of Exercise 7 to obtain the truncation errors occasioned by stopping after five terms of the Maclaurin Series for

(a) $\sin .2$,   (b) $e^{.25}$,   (c) $(1.05)^{2/3}$.

9. How many terms of the Taylor Series about $x - \pi/3$ must be taken to yield (a) $\cos 59°$ correct to six decimals, (b) $\sin 62°$ correct to 6 decimals?

10. Determine the range of x for which the first three terms of (5.10) give a result correct to within .00005 (consider two cases—x positive and x negative).

11. Why can not $\log x$ be expanded in a Maclaurin Series?

12. Try to obtain a Maclaurin Series for the function e^{-1/x^2}.

4. EVALUATION OF INTEGRALS BY FINITE DIFFERENCE METHODS

In the last section, we saw that some integrals could readily be evaluated by the Taylor Series expansion. However, two difficulties may arise; some integrals such as $\int_0^5 \log(1+x^2)\,dx$ involve a range of integration where the series is not valid and so can not be applied; others, such as $\int_0^{100} e^{-t^2}\,dt$, produce a valid series, but one which converges so slowly that it is of no use for practical computation.

Of the various methods which may be employed to circumvent the two difficulties just described, the commonest is

THEOREM 5.3 (Simpson's Rule). If $f(x)$ is a quadratic polynomial, then

(5.11) $$\int_0^2 f(x)\,dx = \tfrac{1}{3}(f_0 + 4f_1 + f_2).$$

Proof. Since $f_x = E^x f_0 = (1+\Delta)^x f_0$, we write

$$\int_0^2 f_x\,dx = \int_0^2 \left[f_0 + x\,\Delta f_0 + \frac{x(x-1)}{2}\Delta^2 f_0 \right] dx$$

$$= \left[x f_0 + \frac{x^2}{2}\Delta f_0 + \left(\frac{x^3}{6} - \frac{x^2}{4}\right)\Delta^2 f_0 \right]_0^2$$

$$= 2f_0 + 2\Delta f_0 + \tfrac{1}{3}\Delta^2 f_0.$$

Now

$$\Delta f_0 = (E-1)f_0 = f_1 - f_0; \qquad \Delta^2 f_0 = (E-1)^2 f_0 = f_2 - 2f_1 + f_0.$$

Hence

$$\int_0^2 f_x\,dx = \tfrac{1}{3}(f_0 + 4f_1 + f_2),$$

as required.

By the change of variable $X = x/h$, we obtain

COROLLARY 1.

$$\int_0^{2h} f_x\,dx = \int_0^2 f_X(h\,dX) = \frac{h}{3}(f_0 + 4f_1 + f_2),$$

where we adopt the usual convention of calling the *consecutive* ordinates used by the names f_0, f_1, f_2 (that is, we write f_0, f_1, f_2, instead of f_0, f_h, f_{2h}).

Thus the area under the curve $y = f(x)$ from 0 to $2h$ is one-third the interval between ordinates multiplied by the sum of the initial ordinate, the final ordinate, and four times the mid-ordinate.

In general, $\int_a^b f(x)\,dx$ represents the area from a to b beneath the curve $y = f(x)$. Now $f(x)$ will not usually be a quadratic function, that is, the graph will not usually be a parabola. However, if we split the interval (a, b) into an even number, $2n$, of parts, then $f(x)$ may be accurately represented by a parabola within each pair of intervals (using different parabolas for different sections of the curve). Then $b = a + 2nh$, and, using Simpson's Rule n times, we get

COROLLARY 2 (the general Simpson Rule).

$$\int_a^{a+2nh} f(x)\, dx = \sum_{r=0}^{n-1} \int_{a+2rh}^{a+2(r+1)h} f(x)\, dx$$

$$= \frac{h}{3} \sum_{r=0}^{n-1} [f_{a+2rh} + 4f_{a+(2r+1)h} + f_{a+(2r+2)h}].$$

Thus,

(5.12) $$\int_a^{a+2nh} f(x)\, dx = \frac{h}{3}[f_0 + 4f_1 + 2f_2 + 4f_3 + \cdots + 2f_{2n-2} + 4f_{2n-1} + f_{2n}],$$

where we again write $f_0, f_1, f_2, \ldots$ rather than $f_a, f_{a+h}, f_{a+2h}, \ldots$

EXAMPLE 5.5. Use Simpson's Rule with two, four, and ten intervals to evaluate

$$\int_1^2 \frac{1}{x}\, dx = \log 2.$$

With two intervals, we obtain

$$\frac{.5}{3}[1 + 2.66667 + .5] = \frac{4.16667}{6} = .69444.$$

With four intervals, we obtain

$$\frac{.25}{3}[1 + 3.2 + 1.33333 + 2.28571 + .5] = \frac{8.31904}{12} = .69325.$$

With ten intervals (using a table of reciprocals),

$$
\begin{array}{rl}
1 + .5 & = \ 1.5 \\
4(.9090909 + .7692308 + .6666667 & = 13.8381580 \\
\quad + .5882353 + .5263158) & \\
2(.8333333 + .7142857 + .625 + .5555556) = & \underline{\ 5.4563492} \\
& 20.7945072
\end{array}
$$

$$\text{Result} = \frac{20.7945072}{30} = .693150.$$

Since $\log 2 = .693147$, we see that the result approaches the true value quite rapidly as the number of intervals is increased.

If we suppose that the given data or the function appearing in $\int_a^b f_x \, dx$ can be represented by a sextic polynomial, then Stirling's formula gives

$$f_x = f_0 + \frac{x}{1!} \mu \delta f_0 + \frac{x^2}{2!} \delta^2 f_0 + \frac{x(x^2 - 1)}{3!} \mu \delta^3 f_0$$

$$+ \frac{x^2(x^2 - 1)}{4!} \delta^4 f_0 + \frac{x(x^2 - 1)(x^2 - 4)}{5!} \mu \delta^5 f_0 + \frac{x^2(x^2 - 1)(x^2 - 4)}{6!} \delta^6 f_0.$$

Integrating from -3 to $+3$, we note that the second, fourth, and sixth terms yield even functions of x, and so do not contribute to the integral. Hence

$$\int_{-3}^3 f_x \, dx = \left[f_0 x + \frac{x^3}{6} \delta^2 f_0 + \frac{1}{4!} \left(\frac{x^5}{5} - \frac{x^3}{3} \right) \delta^4 f_0 \right.$$

$$\left. + \frac{1}{6!} \left(\frac{x^7}{7} - x^5 + \tfrac{4}{3} x^3 \right) \delta^6 f_0 \right]_{-3}^3$$

$$= 6 f_0 + 9 \delta^2 f_0 + \tfrac{33}{10} \delta^4 f_0 + \tfrac{41}{140} \delta^6 f_0.$$

Sixth differences are, in general, very small; so we shall introduce an error of only $\tfrac{1}{140} \delta^6 f_0$ if we replace the coefficient 41 in this expression by the more manageable coefficient 42. Then

$$\int_{-3}^3 f_x \, dx = \tfrac{3}{10} [20 + 30 \delta^2 + 11 \delta^4 + \delta^6] f_0.$$

Simplifying the operator, we use

$$\delta^2 = E - 2 + E^{-1};$$

then

$$20 + \delta^2 [30 + \delta^2(11 + \delta^2)] = (E^3 + E^{-3}) + 5(E^2 + E^{-2}) + (E + E^{-1}) + 6.$$

Hence

$$\int_{-3}^3 f_x \, dx = \tfrac{3}{10} [6 f_0 + (f_1 + f_{-1}) + 5(f_2 + f_{-2}) + (f_3 + f_{-3})].$$

By moving the range three units to the right, this formula can be written as

$$(5.13) \qquad \int_0^6 f(x) \, dx = \tfrac{3}{10} [6 f_3 + (f_2 + f_4) + 5(f_1 + f_5) + (f_0 + f_6)].$$

If the range is 0 to $6h$, we get the most general form of (5.13), known as Weddle's Rule,

(5.14) $\displaystyle\int_0^{6h} f_x\,dx = \frac{3h}{10}\,[6f_3 + (f_2 + f_4) + 5(f_1 + f_5) + (f_0 + f_6)],$

where again we write $f_0, f_1, \dots, f_6$, rather than $f_0; f_h, \dots, f_{6h}$.

The main disadvantage of Weddle's Rule is that the number of intervals must be a multiple of 6.

EXAMPLE 5.6. In an experiment, a quantity G was measured as follows.

$$G(20) = 95.90 \qquad G(24) = \;\;99.56$$
$$G(21) = 96.85 \qquad G(25) = 100.41$$
$$G(22) = 97.77 \qquad G(26) = 101.24$$
$$G(23) = 98.68$$

Compute $\displaystyle\int_{20}^{26} G(x)\,dx$ by Weddle's Rule.

We immediately obtain $\frac{3}{10}(1972.85) = 591.855.$

Had we used Simpson's Rule, the result would have been

$$\tfrac{1}{3}[(f_0 + f_6) + 4(f_1 + f_5) + 2(f_2 + f_4) + 4f_3] = \tfrac{1}{3}(1775.56) = 591.853.$$

We shall conclude this section by developing a formula which does not depend on whether the number of intervals is even (as in Simpson's Rule). Using Bessel's formula in the form (3.16), we can write (with $z = x - \tfrac{1}{2}$)

$$f_x = \mu f_{1/2} + z\,\delta f_{1/2} + \frac{z^2 - \tfrac{1}{4}}{2!}\,\mu\delta^2 f_{1/2} + \frac{z(z^2 - \tfrac{1}{4})}{3!}\,\delta^3 f_{1/2}$$

$$+ \frac{(z^2 - \tfrac{1}{4})(z^2 - \tfrac{9}{4})}{4!}\,\mu\delta^4 f_{1/2} + \frac{z(z^2 - \tfrac{1}{4})(z^2 - \tfrac{9}{4})}{5!}\,\delta^5 f_{1/2}$$

$$+ \frac{(z^2 - \tfrac{1}{4})(z^2 - \tfrac{9}{4})(z^2 - \tfrac{25}{4})}{6!}\,\mu\delta^6 f_{1/2} + \cdots$$

Then

$$\int_0^1 f_x\,dx = \int_{-1/2}^{1/2} f_z\,dz,$$

and the second, fourth, and sixth terms vanish on integration. Hence

$$\int_0^1 f_x\,dx = \left[z\mu f_{1/2} + \frac{1}{2!}\left(\frac{z^3}{3} - \frac{z}{4}\right)\mu\delta^2 f_{1/2} + \frac{1}{4!}\left(\frac{z^5}{5} - \frac{5z^3}{6} + \frac{9z}{16}\right)\mu\delta^4 f_{1/2} \right.$$

$$\left. + \frac{1}{6!}\left(\frac{z^7}{7} - \frac{7z^5}{4} + \frac{259}{48}z^3 - \frac{225}{64}z\right)\mu\delta^6 f_{1/2} + \cdots \right]_{-1/2}^{1/2}$$

$$= \left(\mu - \frac{1}{12}\mu\delta^2 + \frac{11}{720}\mu\delta^4 - \frac{191}{60480}\mu\delta^6 + \cdots\right)f_{1/2}.$$

Thus we get formula

(5.15) $\int_0^1 f_x\,dx = \frac{1}{2}\Big[(f_0 + f_1) - \frac{1}{12}(\delta^2 f_0 + \delta^2 f_1) + \frac{11}{720}(\delta^4 f_0 + \delta^4 f_1) + \cdots\Big].$

As usual, this can be altered to the form

(5.16) $\int_0^h f_x\,dx = \frac{h}{2}\Big[(f_0 + f_1) - \frac{1}{12}(\delta^2 f_0 + \delta^2 f_1) + \cdots\Big].$

If we use (5.16) over n intervals, we obtain

(5.17) $\int_0^{nh} f_x\,dx = \frac{h}{2}\Big[(f_0 + 2f_1 + 2f_2 + \cdots + 2f_{n-1} + f_n)$

$$- \frac{1}{12}(\delta^2 f_0 + 2\delta^2 f_1 + \cdots + 2\delta^2 f_{n-1} + \delta^2 f_n) + \cdots\Big].$$

This last formula simplifies, since we may write

$$\delta^2 f_0 + 2\delta^2 f_1 + \cdots + 2\delta^2 f_{n-1} + \delta^2 f_n$$

$$= \delta^2(1 + 2E + 2E^2 + \cdots + 2E^{n-1} + E^n)f_0$$

$$= \delta^2(E^n - 1)\Big(1 + \frac{2}{E-1}\Big)f_0 = (E^n - 1)\delta^2\Big(\frac{E+1}{E-1}\Big)f_0$$

$$= (E^n - 1)\frac{\Delta^2}{E}\frac{E+1}{\Delta}f_0 = (E^n - 1)\Big(\frac{E^2-1}{E}\Big)f_0$$

$$= 2\mu\delta(E^n - 1)f_0 = 2\mu\delta(f_n - f_0).$$

Hence, the final form is

(5.18) $\int_0^{nh} f_x\,dx = \frac{h}{2}\Big[(f_0 + 2f_1 + \cdots + 2f_{n-1} + f_n)$

$$- \frac{1}{6}(\mu\delta f_n - \mu\delta f_0)$$

$$+ \frac{11}{360}(\mu\delta^3 f_n - \mu\delta^3 f_0)$$

$$- \frac{191}{30240}(\mu\delta^5 f_n - \mu\delta^5 f_0) + \cdots\Big].$$

EXAMPLE 5.7. Use Formula (5.18) to evaluate $\int_0^{.4} \cos x \, dx$. Take $h = .1$. Form the table

x	$f = 10^5 \cos x$	δf	$\delta^2 f$	$\delta^3 f$
$-.2$	98007			
		$+1493$		
$-.1$	99500		-993	
		$+500$		-7
0	100000		-1000	
		-500		$+7$
.1	99500		-993	
		-1493		$+13$
.2	98007		-980	
		-2473		$+25$
.3	95534		-955	
		-3428		$+35$
.4	92106		-920	
		-4348		$+44$
.5	87758		-876	
		-5224		
.6	82534			

Then

$$\int_0^{.4} \cos x \, dx = \frac{1}{20}\left[7.78188 + \frac{.07776}{12} + \frac{11}{720}(.00079) \right]$$

$$= .38942 \qquad (\text{checking, } \sin .4 = .38942).$$

Note that the use of (5.18) presupposes that we can obtain additional functional values so as to give central differences at the beginning and end of the table; if this is not so, a formula like Simpson's rule must be used.

EXERCISES

1. Assuming that fourth and higher differences vanish, prove

$$\int_0^2 f_x \, dx = \tfrac{1}{3}(f_0 + 4f_1 + f_2),$$

that is, Simpson's rule is correct to third differences.

2. Evaluate $\int_0^2 \dfrac{x^3}{e^x - 1}$ by using Simpson's rule with two intervals; with four intervals.

3. $\int_0^1 \dfrac{dx}{1 + x^2} = \dfrac{\pi}{4}$; use Simpson's rule with three and five ordinates to compute an approximation to π. Repeat with nine ordinates.

4. Simpson's rule is called a "closed" formula because it uses functional values f_0 and f_2 at the beginning and end of the range of integration; prove the corresponding "open" formula

$$\int_0^4 f_x \, dx = \tfrac{4}{3}(2f_1 - f_2 + 2f_3).$$

5. Find $\int_{-1.6}^{-1.0} e^x \, dx$ by (a) direct integration, (b) Simpson's rule with six intervals, (c) Weddle's rule, (d) Formula (5.18). Work to six decimals.

6. Evaluate $\int_0^{.5} \cos x^2 \, dx$ by Formula (5.18), and compare with the result obtained by using the Taylor series.

7. A function f_x is experimentally known for four values a, $a + h$, $a + 2h$, $a + 3h$. Making a suitable assumption concerning the form of f_x, prove

$$\int_a^{a+3h} f_x \, dx = \frac{3h}{8} (f_a + 3f_{a+h} + 3f_{a+2h} + f_{a+3h})$$

(convert the range to 0–3).

8. Show that the difference in area between the values of $\int_0^6 f_x \, dx$ obtained by Weddle's rule and Simpson's rule with six intervals is $\tfrac{1}{30}(\Delta^4 + \Delta^5 + \Delta^6)f_0$.

9. Evaluate $\int_0^4 \dfrac{x}{e^x - 1} \, dx$ by (a) Simpson's rule with five ordinates, (b) Taylor's series, (c) Formula (5.18).

10. Suppose that only five ordinates are known; develop the formula

$$\int_0^4 f_x \, dx = \tfrac{2}{45}[7(f_0 + f_4) + 32(f_1 + f_3) + 12f_2].$$

11. For the case of six known ordinates, obtain

$$\int_0^5 f_x \, dx = \tfrac{5}{288}[19(f_0 + f_5) + 75(f_1 + f_4) + 50(f_2 + f_3)].$$

12. Develop the open formula corresponding to Problem 7, that is, express $\int_0^5 f_x \, dx$ in terms of f_1, f_2, f_3, f_4.

13. Compare the assumptions made in Problems 10, 11, and 12.

14. By splitting the interval of integration into n parts and assuming a *linear* approximation to f_x in each interval, develop the trapezoidal rule

$$\int_0^{nh} f_x \, dx = \frac{h}{2}(f_0 + 2f_1 + 2f_2 + 2f_3 + \ldots + 2f_{n-1} + f_n).$$

Use this rule on $\int_1^2 \frac{1}{x} \, dx$ (with ten intervals) and compare the result with that obtained in Example 5.5. On a diagram, shade in the area given by the trapezoidal rule.

15. Assuming that fifth differences of f_x are constant, prove Hardy's Formula

$$\int_0^6 f_x \, dx = .28(f_0 + f_6) + 1.62(f_1 + f_5) + 2.2f_3.$$

Hint: Use the range of integration $(-3, 3)$.

16. The integral

$$\int_0^1 \frac{n \, dn}{n + \alpha e^{-an^2}}$$

arises in chemistry. Make a double-entry table giving the values of this integral for $a = 0, 1, 2, 4$; $\alpha = 0, .25, .5, .75, 1, 2, 3$.

17. The expression

$$\int_0^{\pi/2} \frac{\cos^2 \left(\frac{\pi}{2} \cos \theta \right)}{\sin \theta} \, d\theta$$

occurs in radiation theory (see, for example, E. C. Jordan, *Electromagnetic Waves and Radiating Systems*, p. 318). Change this integral to the form

$$\int_0^1 \frac{\cos^2 \left(\frac{1}{2} \pi x \right)}{1 - x^2} \, dx,$$

and evaluate it by Simpson's rule with two and four intervals (value $= .609$).

18. Use the advancing difference formula to develop the result

$$f'(x) = \left[\Delta + \left(x - \frac{1}{2} \right) \Delta^2 + \left(\frac{x^2}{2} - x + \frac{1}{3} \right) \Delta^3 + \ldots \right] f_0;$$

thence tabulate $G'(x)$ in Example 5.6. (Note that, whereas $\int_a^b G(x) \, dx$ is determined even more exactly than $G(x)$, $G'(x)$ has larger coefficients, and is consequently subject to greater incertitude. Note further that the origin 0 should be moved along at each stage in computing $G'(x)$; this avoids accumulation of error.)

19. Form a difference table for the following experimentally determined function $f(x)$.

x	$f(x)$	x	$f(x)$
1.0	5.47	2.0	33.49
1.1	6.85	2.1	38.85
1.2	8.43	2.2	44.87
1.3	10.25	2.3	51.59
1.4	12.36	2.4	59.06
1.5	14.80	2.5	67.32
1.6	17.62	2.6	76.42
1.7	20.85	2.7	86.39
1.8	24.54	2.8	97.28
1.9	28.74	2.9	109.13
		3.0	121.98

(a) Use Simpson's rule to evaluate $\int_1^3 f(x)\, dx$.

(b) Differentiate the Stirling formula to obtain

$$f'(x) = \left(\mu\delta + x\delta^2 + \frac{3x^2 - 1}{6}\mu\delta^3 \right) f_0.$$

(c) Make a table of the function $f'(x)$ in 1.0(.1)3.0. Note: For the range 1.2 to 2.8 inclusive, we can use part (b) to give $f'(0) = (\mu\delta - \frac{1}{6}\mu\delta^3)f_0$. For the values 1.0 and 1.1 at the beginning of the table, Exercise 18 gives the result $f'(0) = \left(\Delta - \frac{\Delta^2}{2} + \frac{\Delta^3}{3} \right) f_0$ in terms of forward differences. For the values 2.9 and 3.0 at the end of the table, we must develop the results

$$f'(x) = \left[\nabla + (x + \tfrac{1}{2})\nabla^2 + \left(\frac{x^2}{2} + x + \tfrac{1}{3} \right)\nabla^3 \right] f_0 \quad \text{and}$$

$$f'(0) = \left(\nabla + \frac{\nabla^2}{2} + \frac{\nabla^3}{3} \right) f_0$$

in terms of backward differences.

20. Use Simpson's rule with ten intervals to evaluate

$$\int_0^1 \frac{\sin u}{u}\, du$$

to four decimals. Check by using a series to evaluate the integral. Which method would be better for evaluating

$$\int_5^6 \frac{\sin u}{u}\, du?$$

5. ERRORS IN THE FINITE-DIFFERENCE INTEGRATION FORMULAE

In Section 3.5, we considered the error involved in using an interpolating polynomial and found that (unfortunately) it depended upon a high-order derivative. In this section we shall discuss the errors involved in using the integration formulae of the last section and shall find that these errors likewise depend upon high-order derivatives. The operator method we shall use in obtaining these error estimates is due to W. E. Milne (*Numerical Calculus,* p. 108 ff.).

Suppose that the formula under consideration gives an exact result for polynomials of degree m, where m is chosen maximal; for example, $m \geq 3$ for Simpson's rule (cf. Exercise 1, Section 5.4). Then, using Taylor's series with the Lagrangian Remainder (Exercise 7 of Section 5.3), we may write

(5.19)
$$f(x) = \sum_{r=0}^{m} \frac{f^{(r)}(0)}{r!} x^r + \frac{f^{(m+1)}(\theta)}{(m+1)!} x^{m+1}$$

where $0 \leq \theta \leq x$.

We now introduce two linear operators (cf. Exercise 15 of Section 1.2) I and F. I will just be the ordinary definite integral $I = \int_0^a \ldots dx$, where a is the upper limit in the problem; F will represent the operation of the finite-difference formula under consideration. Then, by hypothesis,

$$I \cdot f(x) = F f(x)$$

whenever $f(x)$ is a polynomial of degree not exceeding m. For the general function (5.19), we immediately have

$$(I - F)f(x) = I\left[\frac{f^{(m+1)}(\theta)x^{m+1}}{(m+1)!}\right] - F\left[\frac{f^{(m+1)}(\theta)x^{m+1}}{(m+1)!}\right].$$

But $If(x)$ is the exact value of the integral desired, and $Ff(x)$ is the approximation obtained using the finite-difference operator F; hence $If(x) - Ff(x)$ is just the required expression for the error in $Ff(x)$. We thus obtain

THEOREM 5.4. The error involved in using a finite-difference integration operator F rather than the exact integration operator $I = \int_0^a \ldots dx$ is given by

(5.20)
$$\frac{f^{(m+1)}(\theta)}{(m+1)!}(I - F)x^{m+1},$$

where m is the maximal degree of polynomial for which F is exact, and where θ lies in the range $0 \leq \theta \leq a$.

EXAMPLE 5.8. Obtain the error bound given by (5.20) in the case of Simpson's Rule.

Here we have

$$I f(x) = \int_0^{2h} f(x)\, dx$$

and

$$F f(x) = \frac{h}{3}(f_0 + 4f_1 + f_2).$$

Also we know $m \geq 3$. Trying $m = 3$, we find

$$I x^4 = \int_0^{2h} x^4\, dx = \frac{32}{5} h^5,$$

$$F x^4 = \frac{h}{3}(4h^4 + 16h^4) = \frac{20}{3} h^5.$$

Thus $(I - F)x^4 = \frac{-4}{15} h^5$, and the error is given by

$$-\frac{4}{15} h^5 \frac{f^{(4)}(\theta)}{4!} = -\frac{h^5}{90} f^{(4)}(\theta).$$

In general, we have no information concerning $f^{(4)}(x)$ or about θ; so the best we can do is to state that the absolute value of the error is less than $\frac{h^5}{90} |\hat{f}^{(4)}|$, where $\hat{f}^{(4)}$ is the maximum value of $f^{(4)}(x)$ in the range $0 - 2h$.

If we do not know m, we can find it by trial and error, since we need to know $(I - F)x^{m+1}$ anyway. All we have to do is try the values 1, 2, 3, ... and select the first m for which $(I - F)x^{m+1} \neq 0$.

EXAMPLE 5.9. Use (5.20) to obtain an error bound for Weddle's Rule.

Here

$$I f(x) = \int_0^{6h} f(x)\, dx,$$

$$F f(x) = \frac{3h}{10}[(f_0 + f_6) + 5(f_1 + f_5) + (f_2 + f_4) + 6f_3].$$

Trying 0, 1, 2, ... , we find the following results

$$I x^0 = F x^0 = 6h,$$
$$I x = F x = 18h^2,$$
$$I x^2 = F x^2 = 72h^3,$$
$$I x^3 = F x^3 = 324h^4,$$
$$I x^4 = F x^4 = \frac{7776}{5} h^5,$$
$$I x^5 = F x^5 = 7776h^6,$$
$$I x^6 = \frac{279936}{7} h^7, \quad F x^6 = 39996h^7.$$

Thus, $m + 1 = 6$; we see that Weddle's Rule is exact for polynomials up to and including degree five and that the error, by (5.20), is

$$\frac{f^{(6)}(\theta)}{6!}\left[\frac{279936}{7} - 39996\right]h^7 = -\frac{h^7}{140}f^{(6)}(\theta).$$

Thus, using $\hat{f}^{(6)}$ to denote the maximum value of $f^{(6)}(x)$ in 0–$6h$, we find that the absolute value of the error is bounded by $\dfrac{h^7}{140}|\hat{f}^{(6)}|$.

It is clear from these two examples that the same difficulty in obtaining error bounds exists here as in Section 3.5, namely, the fact that, in general, we know little about the behaviour of high-order derivatives. These may sometimes be estimated by the method described in Section 3.5. However, a more practical approach is simply to halve the interval. If Simpson's Rule with six intervals gives a result of 8.125943 and with twelve intervals gives a result of 8.125932, then we should be safe in stating that the result (to four decimals) is 8.1259 (and is quite likely even 8.125932). This concrete test of increasing the number of intervals and actually seeing the finite-difference approximation approach its limiting value is far more fruitful than attempting to estimate high-order derivatives.

EXERCISES

1. (All references are to the Exercises following Section 5.4.) Use (5.20) to obtain the error in
   (a) the open Simpson's Rule (Exercise 4),
   (b) the Simpson three-eighths Rule (Exercise 7),
   (c) the $\frac{2}{45}$ths Rule (Exercise 10),
   (d) the $\frac{5}{288}$ths Rule (Exercise 11),
   (e) Hardy's Formula (Exercise 15),
   (f) the trapezoidal Rule (Exercise 14).

2. Working to five decimals, use Simpson's Rule to evaluate
$$\int_1^3 \log_{10} x \, dx.$$
   Use (a) two intervals,   (b) four intervals,
        (c) eight intervals,   (d) sixteen intervals.
   Note that the goodness of the result can be estimated in this manner without using high-order derivatives.

3. Repeat Exercise 2 for $\displaystyle\int_5^9 \log_{10} x \, dx$.

4. Obtain error estimates, using (5.20), for Dufton's Rule (eleven ordinates)
$$\int_0^{10h} f(x) \, dx = \frac{5h}{2}(f_1 + f_4 + f_6 + f_9).$$

6. THE EULER-MACLAURIN SERIES

The formal expansion

(5.21)
$$\frac{x}{e^x - 1} = \sum_{r=0}^{\infty} \frac{B_r}{r!} x^r$$

is of considerable importance; the coefficients B_r are known as the Bernoulli numbers and can be found, by cross-multiplying and comparing coefficients, as

$$B_0 = 1, \quad B_1 = -\tfrac{1}{2}, \quad B_2 = \tfrac{1}{6}, \quad B_3 = 0, \quad B_4 = -\tfrac{1}{30}, \quad B_5 = 0,$$

$$B_6 = \tfrac{1}{42}, \quad B_7 = 0, \quad B_8 = -\tfrac{1}{30}, \quad B_9 = 0, \quad B_{10} = \tfrac{5}{66}, \ \ldots$$

We shall use the expansion (5.21) in formally obtaining a connection between

$$\int_0^n f(x)\, dx \quad \text{and} \quad \sum_{x=0}^n f(x).$$

Suppose $F(x)$ is a function such that $\Delta F(x) = f(x)$. Then

(5.22)
$$F(x+1) - F(x) = f(x).$$

Summing (5.22) as x ranges from 0 to $n - 1$, we obtain

(5.23)
$$F(n) - F(0) = \sum_{x=0}^{n-1} f(x).$$

Now, by the Taylor series (5.4),

$$f(x+1) = \sum_{r=0}^{\infty} \frac{D^r f(x)}{r!} = \left[\sum_{r=0}^{\infty} \frac{D^r}{r!} \right] f(x) = e^D f(x).$$

But
$$f(x+1) = E f(x) = (1 + \Delta) f(x).$$

Hence, comparing the operators,

(5.24)
$$e^D = 1 + \Delta.$$

Using (5.24), we have

$$F(x) = \Delta^{-1} f(x) = \frac{1}{e^D - 1} f(x)$$

$$= D^{-1} \frac{D}{e^D - 1} f(x)$$

$$= D^{-1} \sum_{r=0}^{\infty} \frac{B_r D^r}{r!} f(x)$$

$$= \left[D^{-1} - \tfrac{1}{2} + \tfrac{1}{12} D - \tfrac{1}{720} D^3 + \tfrac{1}{30240} D^5 - \ldots \right] f(x)$$

$$= D^{-1} f(x) - \tfrac{1}{2} f(x) + \tfrac{1}{12} f'(x) - \tfrac{1}{720} f'''(x) + \ldots$$

Evaluating this expression for $x = 0$ and $x = n$ yields

(5.25) $\quad F(n) - F(0) = \int_0^n f(x)\,dx - \frac{1}{2}[f(n) - f(0)]$

$$+ \frac{1}{12}[f'(n) - f'(0)] - \frac{1}{720}[f'''(n) - f'''(0)] + \ldots$$

A comparison of (5.23) with (5.25) then gives the Euler-Maclaurin Formula

(5.26) $\quad \sum_{x=0}^n f(x) = \int_0^n f(x)\,dx + \frac{1}{2}[f(n) + f(0)]$

$$+ \frac{1}{12}[f'(n) - f'(0)] - \frac{1}{720}[f'''(n) - f'''(0)] + \ldots$$

This relation is normally more useful for summing series than for evaluating integrals; the truncation error on the right-hand side will, in general, be less than the first term omitted (we shall return to this point in Section 8).

EXAMPLE 5.10. Compute $\sum_{x=1}^{\infty} x^{-3}$ correct to seven decimals.

In order to ensure rapid diminution in the terms, we add up the terms for $x = 1, \ldots , 9$ separately.

$$\sum_{x=1}^{9} x^{-3} = 1.125 + .03703704 + .015625 + .008 + .00462963$$

$$+ .00291545 + .00195313 + .00137174 = 1.19653199.$$

In this problem,

$$f(x) = \frac{1}{x^3}, \quad f'(x) = \frac{-3}{x^4}, \quad f'''(x) = \frac{-60}{x^6}, \quad f^{v}(x) = \frac{-2520}{x^8}.$$

Hence

$$\sum_{10}^{\infty} f(x) = \int_{10}^{\infty} \frac{1}{x^3}\,dx + \frac{1}{2}f(10) - \frac{1}{12}f'(10) + \frac{1}{720}f'''(10) - \frac{1}{30240}f^{v}(10) + \ldots$$

$$= .005 + .0005 + .000025 - .00000008 + .00000000$$

$$= .00552492.$$

Our final result is

$$1.19653199 + .00552492 = 1.2020569,$$

rounding off the guard figure.

EXERCISES

1. Use the Euler-Maclaurin expansion to prove

(a) $\displaystyle\sum_{x=1}^{n} x^2 = \frac{n(n + 1)(2n + 1)}{6}$,

(b) $\displaystyle\sum_{x=1}^{n} x^3 = \frac{n^2(n + 1)^2}{4}$.

2. Use the Euler-Maclaurin expansion to compute

$$f(n) = \sum_{r=1}^{n} \frac{1}{r} - \log n$$

for (a) $n = 10^5$, (b) $n = 10^8$.
(The value of $\gamma = \lim_{n \to \infty} f(n) = .577215665$.)

3. Use the identity

$$\frac{\pi^2}{6} = \sum_{n=1}^{\infty} \frac{1}{n^2}$$

to compute π^2 to eight decimals.

4. Verify numerically that $\sum_{n=1}^{\infty} n^{-4} = \pi^4/90$.

5. Use the Euler-Maclaurin expansion to evaluate $\sum_{50}^{100} x^{-1}$.

6. (a) Putting $z = 2ix$ in the identity

$$\cot x = i \left[\frac{e^{ix} + e^{-ix}}{e^{ix} - e^{-ix}} \right],$$

deduce that

$$x \cot x = \frac{z}{2} + \frac{z}{e^z - 1}$$

$$= \sum_{r=0}^{\infty} (-)^r \frac{B_{2r}(2x)^{2r}}{(2r)!} .$$

(b) Assuming the analytic relation

$$x \cot x = 1 + \sum_{n=1}^{\infty} \frac{2x^2}{x^2 - n^2\pi^2} ,$$

use the binomial expansion and (a) to show that

$$\sum_{n=1}^{\infty} \frac{1}{n^{2r}} = (-)^{r-1} \frac{(2\pi)^{2r} B_{2r}}{2(2r)!} .$$

7. Generalize the Euler-Maclaurin formula to the case

$$f(a) + f(a + h) + \ldots + f(a + nh)$$

$$= \frac{1}{h} \int_a^{a+nh} f(x)\, dx + \tfrac{1}{2}[f(a) + f(a + nh)]$$

$$+ \frac{h}{12}[f'(a + nh) - f'(a)] - \frac{h^3}{720}[f'''(a + nh) - f'''(a)] + \ldots$$

8. Use the result of Exercise 7 to compute $\int_1^3 \sqrt{x^3 + 10}\, dx$ with $h = 1/2$; compare with the result obtained from Simpson's rule with $h = 1/2$.

9. From Exercise 4, we have

$$\sum_{r=1}^{n} \frac{1}{r} \doteq \log n + \gamma;$$

deduce that

$$\sum_{100}^{200} \frac{1}{r} \doteq \log 2.$$

Compute a more exact value of $\sum_{100}^{200} \frac{1}{r}$ by the Euler-Maclaurin series.

7. THE EULER TRANSFORMATION

Suppose we have an alternating series

$$S = u_1 - u_2 + u_3 - u_4 + \dots$$

where

$$u_n > 0, \quad u_n > u_{n+1}, \quad u_n \to 0 \quad \text{as} \quad n \to \infty.$$

Suppose further that this series converges very slowly; then, in its given form, it is of little use for computational purposes. However, we can evaluate S if we form a difference table of the values u_i (note that the u_i will all be positive). Such a table assumes the following appearance.

$$
\begin{array}{ccccc}
u_1 \\
 & \Delta u_1 \\
u_2 & & \Delta^2 u_1 \\
 & \Delta u_2 & & \Delta^3 u_1 \\
u_3 & & \Delta^2 u_2 \\
 & \Delta u_3 \\
u_4
\end{array}
$$

If we use Δ and E to refer to operations on the (positive) entries u_i in this table, we have

$$
\begin{aligned}
S &= u_1 - E u_1 + E^2 u_1 - E^3 u_1 + \dots \\
&= (1 - E + E^2 - E^3 + \dots) u_1 \\
&= (1 + E)^{-1} u_1 \\
&= (2 + \Delta)^{-1} u_1 \\
&= \frac{1}{2}\left(1 + \frac{\Delta}{2}\right)^{-1} u_1 \\
&= \frac{1}{2}\left[1 - \frac{\Delta}{2} + \frac{\Delta^2}{4} - \frac{\Delta^3}{8} + \frac{\Delta^4}{16} - \dots\right] u_1.
\end{aligned}
$$

This formula

(5.27) $$u_1 - u_2 + u_3 - \dots = \frac{1}{2}\left[u_1 - \frac{\Delta u_1}{2} + \frac{\Delta^2 u_1}{4} - \dots\right]$$

which expresses the sum of the series in terms of the first term u_1 and its leading differences, is called the Euler transformation.

EXAMPLE 5.11. Compute log 2 to five decimals from the series

$$1 - \tfrac{1}{2} + \tfrac{1}{3} - \tfrac{1}{4} + \cdots$$

In order to have the differences smaller, we add up the first ten terms separately. For the terms after the tenth, we form the following difference table.

x	$u_x = \dfrac{10^6}{x}$	Δu	$\Delta^2 u$	$\Delta^3 u$	$\Delta^4 u$	$\Delta^5 u$	$\Delta^6 u$
11	90909						
		−7576					
12	83333		1166				
		−6410		−251			
13	76923		915		70		
		−5495		−181		−29	
14	71428		734		41		25
		−4761		−140		−4	
15	66667		594		37		
		−4167		−103			
16	62500		491				
		−3676					
17	58824						

We have

$$\sum_{x=1}^{10} \frac{(-1)^{x+1}}{x} = .645634,$$

and

$$\sum_{x=10}^{\infty} \frac{(-1)^{x+1}}{x} = \frac{1}{2}\left[.090909 + \frac{.007576}{2} + \frac{.001166}{4} + \frac{.000251}{8} + \frac{.000070}{16} + \cdots \right]$$

$$= \; = .047513.$$

The sum of the series is then .693147, the last figure being a guard (it happens to be correct here). We might note that, in order to obtain the value .69315 from the series in its original form, it would have been necessary to add more than 10^5 terms (and such an addition would involve a considerable accumulation of round-off errors!).

EXERCISES

1. Assuming

$$\frac{1}{1^2} + \frac{1}{2^2} + \frac{1}{3^2} + \cdots = \frac{\pi^2}{6},$$

deduce that

$$\frac{1}{1^2} + \frac{1}{3^2} + \frac{1}{5^2} + \cdots = \frac{\pi^2}{8} \quad \text{and} \quad \frac{1}{1^2} - \frac{1}{2^2} + \frac{1}{3^2} - \frac{1}{4^2} + \cdots = \frac{\pi^2}{12}.$$

Use this second series, with the Euler transformation, to compute $\pi^2/12$ to six decimals.

2. (a) Let

$$S = \frac{1}{1^3} - \frac{1}{3^3} + \frac{1}{5^3} - \frac{1}{7^3} + \cdots ;$$

use the Euler transformation and a difference table constructed from the terms $1/9^3, \ldots, 1/21^3$, to compute S to six decimals. Check that $S = \pi^3/32$.

(b) How many terms of the series in (a) would be necessary in order to obtain the same result by simply adding terms?

3. Use the Euler transformation to evaluate

(a) $\displaystyle\sum_{x=1}^{\infty} \frac{(-)^{x+1}}{\sqrt{x}}$ (to four decimals),

(b) $1 - \dfrac{1}{2^3} + \dfrac{1}{3^3} - \dfrac{1}{4^3} + \cdots$ (to six decimals).

4. Compute

$$\sum_{x=1}^{1000} \frac{(-)^{x+1}}{x} \quad \text{and} \quad \sum_{x=1}^{1000} \frac{(-)^{x+1}}{x^2}.$$

8. ASYMPTOTIC EXPANSIONS

Since the concept of an asymptotic series is quite different from that of an ordinary convergent or divergent series, it will be well to commence with an example. Most asymptotic series encountered in elementary work are generated by integration by parts. While a formal definition of an asymptotic series is given in Exercise 7 at the end of this section, we can give a rough-and-ready initial definition as

"An asymptotic series for a function $f(x)$ is an expression

$$f(x) = S_n(x) + R_n(x)$$

where $S_n(x)$ is a finite series of n terms and $R_n(x)$ is a term such that

$$\operatorname*{Lim}_{n \to \infty} R_n(x) = \infty \quad (x \text{ fixed}),$$

but

$$\operatorname*{Lim}_{x \to \infty} R_n(x) = 0 \quad (n \text{ fixed})".$$

EXAMPLE 5.12. Evaluate $\displaystyle\int_x^{\infty} e^{-t^2/2} \, dt$, given that x is large.

We have

$$\int_x^{\infty} e^{-t^2/2} \, dt = \int_x^{\infty} -\frac{1}{t} (-t e^{-t^2/2}) \, dt$$

$$= \left[-\frac{1}{t} e^{-t^2/2} - \frac{1}{t^2} \int e^{-t^2/2} \, dt \right]_x^{\infty}$$

$$= \frac{1}{x} e^{-x^2/2} - \int_x^{\infty} \frac{1}{t^2} e^{-t^2/2} \, dt.$$

Treating the second integral in the same manner,

$$\int_x^\infty \frac{1}{t^2} e^{-t^2/2}\, dt = \int_x^\infty -\frac{1}{t^3}(-te^{-t^2/2})\, dt$$

$$= \frac{1}{x^3} e^{-x^2/2} - \int_x^\infty \frac{3}{t^4} e^{-t^2/2}\, dt.$$

Hence

$$\int_x^\infty e^{-t^2/2}\, dt = \frac{1}{x} e^{-x^2/2} - \frac{1}{x^3} e^{-x^2/2} + \int_x^\infty \frac{3}{t^4} e^{-t^2/2}\, dt.$$

Continuing in this fashion, we ultimately obtain the expression

$$\int_x^\infty e^{-t^2/2}\, dt = e^{-x^2/2}\left[\frac{1}{x} - \frac{1}{x^3} + \frac{1\cdot 3}{x^5} - \frac{1\cdot 3\cdot 5}{x^7} + \cdots \right.$$
$$\left. + (-)^{n-1}\frac{1\cdot 3\cdot 5\ldots(2n-3)}{x^{2n-1}}\right] + (-)^n R_n,$$

where

$$R_n = 1\cdot 3\cdot 5\ldots(2n-1)\int_x^\infty e^{-t^2/2}\frac{1}{t^{2n}}\, dt.$$

If we think of continuing the process indefinitely, we can write *formally*

$$\int_x^\infty e^{-t^2/2}\, dt \sim e^{-x^2/2}\left[\frac{1}{x} - \frac{1}{x^3} + \frac{1\cdot 3}{x^5} - \frac{1\cdot 3\cdot 5}{x^7} + \cdots\right].$$

This is read as "$\int_x^\infty e^{-t^2/2}\, dt$ is asymptotically equal to ...". It should be stressed that it means exactly the same as the closed finite expression involving R_n; it does *not* imply convergence (in fact, the series is not convergent). The best way of regarding the matter is to consider that the asymptotic series stands for a whole infinitude of finite series, namely,

$$\int_x^\infty e^{-t^2/2}\, dt = e^{-x^2/2}\left[\frac{1}{x}\right] - R_1$$

$$= e^{-x^2/2}\left[\frac{1}{x} - \frac{1}{x^3}\right] + R_2$$

$$= e^{-x^2/2}\left[\frac{1}{x} - \frac{1}{x^3} + \frac{3}{x^5}\right] - R_3$$

$$= \ldots$$

Clearly, we have a great deal of liberty in selecting which series we use; the crucial matter is the behaviour of R_n. Now

$$R_n = 1\cdot 3\cdot 5\ldots(2n-1)\int_x^\infty e^{-t^2/2}\frac{1}{t^{2n}}\, dt$$

$$= \frac{1\cdot 3\cdot 5\ldots(2n-1)}{x^{2n+1}} e^{-x^2/2} - R_{n+1}.$$

Since both R_n and R_{n+1} are positive (being integrals of positive functions), we see that the error term R_n is less than the first term of the asymptotic series which we neglect. This allows us to decide how many terms should be taken in the asymptotic series; clearly R_n starts off by decreasing, and then ultimately becomes very large. So we should stop just before the smallest term. For instance, if $x = 3$, we get

$$\int_3^\infty e^{-t^2/2}\,dt \sim e^{-4.5}\left[\frac{1}{3} - \frac{1}{3^3} + \frac{1\cdot 3}{3^5} - \frac{1\cdot 3\cdot 5}{3^7} + \cdots\right].$$

A minimal bound on R_n is given by $R_4 < e^{-4.5}\dfrac{1\cdot 3\cdot 5\cdot 7}{3^9}$; hence, we select the particular finite series

$$\int_3^\infty e^{-t^2/2}\,dt = e^{-4.5}\left[\frac{1}{3} - \frac{1}{3^3} + \frac{1\cdot 3}{3^5} - \frac{1\cdot 3\cdot 5}{3^7}\right] + R_4$$

$$= .3018e^{-4.5} + R_4,$$

where $R_4 < .0053e^{-4.5}$.

Hence we may write

$$.3018e^{-4.5} < \int_3^\infty e^{-t^2/2}\,dt < .3071e^{-4.5}.$$

Clearly, if x is large, the terms of the asymptotic series will become very tiny before they start to increase; so one will be able to get an excellent value. However, it must be noted that, whereas *in a convergent series one can always theoretically get an answer to any arbitrary degree of accuracy by taking enough terms, in an asymptotic series there is a bound on the accuracy possible.* Once x is given, our answer can not be improved past a certain limit which is determined by the minimum value of R_n.

While further examples are given in the exercises, it should be noted that (contrary to statements in some texts) the Euler-Maclaurin expansion usually provides one with an asymptotic series (it is only rarely that the expansion converges).

EXERCISES

1. Use the series just obtained to evaluate $\int_4^\infty e^{-\frac{1}{2}t^2}\,dt$ as accurately as possible from the asymptotic series. Give the answer and the error estimate in the form Ke^{-8}.

2. How large must x be in order that the value

$$\int_x^\infty \frac{1}{\sqrt{2\pi}} e^{-t^2/2}\,dt \sim \frac{e^{-x^2/2}}{x\sqrt{2\pi}}$$

be correct to three figures?

3. Show that

$$\int_x^\infty \frac{e^{x-t}}{t}\,dt = \frac{1}{x} - \frac{1!}{x^2} + \frac{2!}{x^3} - \frac{3!}{x^4} + \dots + (-)^{n-1}\frac{(n-1)!}{x^n} + (-)^n R_n,$$

where $R_n < \dfrac{n!}{x^{n+1}}$.

Use the corresponding asymptotic series

$$\int_x^\infty \frac{e^{x-t}}{t}\,dt \sim \frac{1}{x} - \frac{1!}{x^2} + \frac{2!}{x^3} - \dots$$

to compute $\int_4^\infty \dfrac{dt}{te^t}$ as accurately as possible. Compute also $\int_6^\infty \dfrac{dt}{te^t}$.

4. In the theory of diffraction at a straight edge, one meets the Cornu integrals

$$X(a) = \int_0^a \cos\left(\tfrac{1}{2}\pi v^2\right)dv, \qquad Y(a) = \int_0^a \sin\left(\tfrac{1}{2}\pi v^2\right)dv,$$

$$R(a) = \int_a^\infty \cos\left(\tfrac{1}{2}\pi v^2\right)dv, \qquad S(a) = \int_a^\infty \sin\left(\tfrac{1}{2}\pi v^2\right)dv.$$

(a) Assuming $\int_0^\infty e^{-a^2v^2}\,dv = \dfrac{\sqrt{\pi}}{2a}$, prove

$$X(\infty) = Y(\infty) = \tfrac{1}{2}.$$

Hint: Consider $X(\infty) + i\,Y(\infty)$, and use

$$e^{iz} = \cos z + i\sin z;$$

you will also need the fact that $\left\{\dfrac{1-i}{\sqrt2}\right\}^2 = -i$.

(b) Write down the first three terms of a power series for $X(a)$; for what values of a would the first two terms give $X(a)$ correct to four decimals?

(c) Prove

$$R(a) \sim \cos\frac{\pi a^2}{2}\left[\frac{1}{\pi^2 a^3} - \frac{1\cdot3\cdot5}{\pi^4 a^7} + \frac{1\cdot3\cdot5\cdot7\cdot9}{\pi^6 a^{11}} - \dots\right]$$
$$- \sin\frac{\pi a^2}{2}\left[\frac{1}{\pi a} - \frac{1\cdot3}{\pi^3 a^5} + \frac{1\cdot3\cdot5\cdot7}{\pi^5 a^9} - \dots\right],$$

and use the result to obtain $R(10)$ correct to six decimals.

(d) Find $X(10)$ correct to six decimals.

(e) Obtain an asymptotic expansion for $S(a)$.

5. Apply the Euler-Maclaurin formula with $f(x) = \log x$ to obtain

$$\log x! = \sum_{k=1}^x \log x \sim (x + \tfrac{1}{2})\log x - x + C + \sum_{r=1}^\infty \frac{B_{2r}}{2r(2r-1)}\frac{1}{x^{2r-1}}$$

where $C \sim 1 - \sum_{r=1}^\infty \dfrac{B_{2r}}{2r(2r-1)}$. (It can be shown by advanced methods that C "sums" to .919 $= \tfrac{1}{2}\log 2\pi$.)

6. Use the formula in Exercise 5 (Stirling's Formula) to compute log 20! (compare with $\log_{10} 20! = 18.38612$).

7. Poincaré's definition of an asymptotic series for a function $f(x)$ is a divergent series $\sum\limits_{r=0}^{\infty} a_r x^{-r}$ such that

$$\mathrm{Lim}_{x \to \infty} \left[\sum_{r=0}^{n} a_r x^{-r} - f(x) \right] x^n = 0 \qquad (n \text{ fixed}),$$

$$\mathrm{Lim}_{n \to \infty} \left[\sum_{r=0}^{n} a_r x^{-r} - f(x) \right] x^n = \infty \qquad (x \text{ fixed}).$$

Show that the expansion obtained in Exercise 3 satisfies this definition of an asymptotic expansion.

9. THE LAGRANGE SERIES

Occasionally one needs to obtain a series from an implicit relationship between x and y of the form

(5.28) $x = y\, f(x),$

where $f(x)$ may be expanded as a convergent series in x,

(5.29) $f(x) = f_0 + f_1 x + f_2 x^2 + \dots,$

with constant term $f_0 \neq 0$. The problem then is to solve for x as a power series in y, that is, to find constants $a_0, a_1, a_2, \dots$ such that

(5.30) $x = a_0 + a_1 y + a_2 y^2 + a_3 y^3 + \dots;$

since it is clear from (5.28) that $x = 0$ when $y = 0$, we must have $a_0 = 0$. Furthermore,

$$y = x[f(x)]^{-1} = b_1 x + b_2 x^2 + b_3 x^3 + \dots$$

where the constants b_i could be determined, although they will not be needed.

To determine the constants a_i, we differentiate (5.30) and obtain

$$1 = [a_1 + 2a_2 y + 3a_3 y^2 + \dots]\frac{dy}{dx}$$

$$= \sum_{r=1}^{\infty} r a_r y^{r-1} \frac{dy}{dx} ;$$

multiply through by $[f(x)]^n$, and obtain

(5.31) $[f(x)]^n = \sum\limits_{r=1}^{\infty} r a_r [f(x)]^n y^{r-1} \dfrac{dy}{dx}.$

The identity (5.31) represents two ways of writing the power series for $[f(x)]^n$. We now evaluate

$$\frac{d^{n-1}}{dx^{n-1}} [f(x)]^n \Big|_{x=0}$$

from the right-hand side of (5.31). This quantity will be

$$(n-1)!\left[\text{coefficient of } x^{n-1} \text{ in } \sum_{r=1}^{\infty} ra_r x^n y^{r-n-1} \frac{dy}{dx}\right].$$

For $r = n$, the term

$$ra_r x^n y^{r-n-1} \frac{dy}{dx} = na_n x^n \frac{1}{y} \frac{dy}{dx}$$

$$= na_n x^n \frac{b_1 + 2b_2 x + \dots}{b_1 x + b_2 x^2 + \dots}$$

$$= na_n x^n \frac{1}{x} [1 + B_1 x + \dots],$$

and the term in x^{n-1} is $na_n x^{n-1}$.

For $r \neq n$, the term

$$ra_r x^n y^{r-n-1} \frac{dy}{dx} = ra_r x^n \frac{1}{r-n} \frac{d}{dx} [y^{r-n}].$$

Now no power of x, either negative or positive, has derivative x^{-1}; consequently, the derivative of y^{r-n} does not contain x^{-1}. Therefore, for $r \neq n$, the expression

$$ra_r x^n y^{r-n-1} \frac{dy}{dx}$$

contains no term in x^{n-1}. Thus,

(5.32) $$\sum_{r=1}^{\infty} ra_r x^n y^{r-n-1} \frac{dy}{dx} = \dots + na_n x^{n-1} + \dots$$

Differentiation of (5.31), and comparison with (5.32), yields

(5.33) $$\frac{d^{n-1}}{dx^{n-1}} [f(x)]^n \bigg|_{x=0} = (n-1)!na_n = n!a_n.$$

Formula (5.33) gives us the so-called Lagrange series (5.30) for x in terms of y. Once having obtained the series formally, we may then discuss its convergence.

EXAMPLE 5.13: Express x as a series in y, given that

$$x = ye^{-x}.$$

Here $[f(x)]^n = e^{-nx}$; hence

$$\frac{d^{n-1}}{dx^{n-1}} [f(x)]^n = (-n)^{n-1} e^{-nx},$$

and the Lagrange coefficients are

$$a_n = \frac{1}{n!} (-n)^{n-1} e^0 = \frac{(-n)^{n-1}}{n!}.$$

Hence
$$x = \sum_{n=1}^{\infty} \frac{(-n)^{n-1}}{n!} y^n.$$

Forming the Cauchy ratio,

$$R = \lim_{n \to \infty} \left| \frac{a_{n+1} y^{n+1}}{a_n y^n} \right|$$

$$= \lim_{n \to \infty} \left| \left(1 + \frac{1}{n}\right)^{n-1} y \right|$$

$$= \lim_{n \to \infty} \left| \left(1 + \frac{1}{n}\right)^{n} \left(1 + \frac{1}{n}\right)^{-1} y \right| = |ey|,$$

we see that the series obtained converges for $|y| < e^{-1}$.

EXAMPLE 5.14: J. O. Irwin [Biometrics, Vol. 15, No. 2 (1959), pp. 324–6] required the value of x, given $y = 3.02$, from the equation

$$x = y - ye^{-x}.$$

Substitute $x = X + y$; and obtain

$$X = -ye^{-X-y} = (-ye^{-y})e^{-X}.$$

Now substitute $(-ye^{-y}) = Y$, and we have the equation in the form

$$X = Ye^{-X}$$

solved in Example 5.13. Thus

$$X = \sum_{n=1}^{\infty} \frac{(-n)^{n-1}}{n!} Y^n;$$

$$x + y = - \sum_{n=1}^{\infty} \frac{n^{n-1}}{n!} (ye^{-y})^n;$$

$$x = y - \sum_{n=1}^{\infty} \frac{n^{n-1}}{n!} (ye^{-y})^n.$$

This last series gives x in terms of y and will converge if $ye^{-y} < e^{-1}$, that is, if $y > 1$. Consequently, Irwin's value 3.02 may be used to give

$$ye^{-y} = 3.02e^{-3.02} = .1471769,$$

$$x = 3.02 - \sum_{n=1}^{\infty} \frac{n^{n-1}}{n!} (.1471769)^n.$$

Convergence is rapid, and we quickly find, using six terms of the series, that

$$x = 2.8447.$$

EXERCISES

1. Use the Lagrange formula to find x in terms of y, given that $x = y(1 - x)$. Check your answer by solving algebraically for x to give $x = y/(1 + y)$.

2. Use the Lagrange formula to find x in terms of y, given that $x = y(1 - x^2)$. Check by solving algebraically to find

$$x = \frac{1}{2y}[-1 \pm \sqrt{1 + 4y^2}].$$

Note that the Lagrange formula produces a single solution, namely, that solution having $x = y = 0$.

3. Apply the Lagrange formula in the following cases:

   (a) $x = y(1 + x^3)$,
   (b) $x = ye^{-x^2}$,
   (c) $x = y(1 + x^2)^{-1}$,
   (d) $x = y(1 - \sin x)$,
   (e) $x = y \cos x$,
   (f) $x = y(e^x + e^{-x})$,
   (g) $x = y(1 + \log x)$.

4. Apply the Lagrange formula, with a suitable transformation, to obtain x in terms of y, given $x = ay + bye^x$.

5. Generalize the Lagrange formula to the case when, given $x = yf(x)$, it is required to express a function $g(x)$ in the form

$$g(x) = \sum_{r=0}^{\infty} a_r y^r.$$

Show that

$$a_n = \frac{1}{n!} \frac{d^{n-1}}{dx^{n-1}} \{g'(x)[f(x)]^n\} \Big|_{x=0}.$$

6. Apply the result of Exercise 5 to expand e^{2x} as a series in $y = xe^x$. [Write $x = ye^{-x}$.]

CHAPTER SIX

Numerical Solution of Differential Equations

1. NATURE OF THE PROBLEM

Any equation involving $x, y, dy/dx, \ldots, d^n y/dx^n$ is called a differential equation of the nth order; it is well known that, under rather general conditions, the solution is of the form $y = g(x, c_1, \ldots, c_n)$, where the c_i are arbitrary constants. We shall exemplify the methods available for solving such equations numerically by considering first-order equations (largely); we shall also restrict ourselves to first-degree equations (that is, equations in which dy/dx appears to the first power only). Thus our typical equation will be

(6.1)
$$\frac{dy}{dx} = f(x, y)$$

and the general solution will be

(6.2)
$$y = g(x, c).$$

Since this solution involves an arbitrary constant, we can prescribe one initial condition; we shall suppose that this is the initial value $x = a$, $y = b$.

EXAMPLE 6.1.

$$10x \frac{dy}{dx} + y^2 - 2 = 0; \qquad (x = 4, y = 1).$$

Separating variables:

$$\frac{dy}{2 - y^2} = \frac{dx}{10x},$$

$$\frac{1}{2\sqrt{2}} \log \frac{\sqrt{2} + y}{\sqrt{2} - y} = \frac{1}{10} \log x + \log c,$$

$$\log \frac{\sqrt{2} + y}{\sqrt{2} - y} = \frac{\sqrt{2}}{5} \log x + 2\sqrt{2} \log c,$$

$$\frac{\sqrt{2} + y}{\sqrt{2} - y} = kx^{\sqrt{2}/5}.$$

The general solution then is

$$y = \frac{\sqrt{2}(kx^{\sqrt{2}/5} - 1)}{kx^{\sqrt{2}/5} + 1}.$$

However, the general solution represents an infinite family of plane curves; we want that particular curve passing through the point (4, 1). Hence we obtain k from the equation

$$\frac{\sqrt{2} + 1}{\sqrt{2} - 1} = k4^{\sqrt{2}/5};$$

substituting, our required solution is

$$y = \sqrt{2}\left[\frac{x^{\sqrt{2}/5}(\sqrt{2} + 1) - 4^{\sqrt{2}/5}(\sqrt{2} - 1)}{x^{\sqrt{2}/5}(\sqrt{2} + 1) + 4^{\sqrt{2}/5}(\sqrt{2} - 1)}\right].$$

The preceding example illustrates the fact that the exact analytic solution of a problem may not be of much use for numerical computation; if, for example, we wanted to tabulate y from 4(.1)5, we should not wish to use the solution just obtained. And, of course, there are many problems for which no exact analytic solution is available. Our requirements, then, are methods for tabulating a solution of (6.1), given an initial value $x = a$, $y = b$.

EXERCISES

1. Find the exact analytic solution of

$$\frac{dy}{dx} = \frac{x + y}{x - y}$$

passing through (3, −1). (In such a homogeneous equation, solve by substituting $y = vx$.) Compute y for $x = 3.2$.

2. Verify that

$$y' = 1 - 2xy$$

has solution

$$y = e^{-x^2} \int_0^x e^{t^2}\, dt.$$

3. In Example 6.1, compute y for $x = 4.2$.

2. SOLUTION IN SERIES

The method of this section is usually not convenient unless the equation has the form

(6.3) $$\sum_{r=0}^{n} A_i(x) D^r y = 0.$$

However, for equations of this kind it is so important that we discuss it even though (6.3) is, in general, of higher order than the type (6.1) with which we are mainly concerned.

EXAMPLE 6.2.

$$x^2 \frac{d^2y}{dx^2} + x \frac{dy}{dx} + \left(x^2 - \frac{1}{4}\right)y = 0.$$

Let $y = x^m \sum_{s=0}^{\infty} a_s x^s$ with $a_0 \neq 0$; then

$$\frac{dy}{dx} = \sum_{s=0}^{\infty} (m + s)a_s x^{m+s-1}, \qquad \frac{d^2y}{dx^2} = \sum_{s=0}^{\infty} (m + s)(m + s - 1)a_s x^{m+s-2}.$$

Substituting in the given equation, we find

$$\sum_{s=0}^{\infty} \left[(m + s)(m + s - 1)a_s + (m + s)a_s + \left(x^2 - \frac{1}{4}\right)a_s \right] x^{m+s} = 0,$$

that is,

$$\sum_{s=0}^{\infty} \left[(m + s)^2 a_s + a_{s-2} - \frac{1}{4} a_s \right] x^{m+s} = 0,$$

with the convention $a_{-2} = a_{-1} = 0$.

We thus get the recurrence relation

$$\left[(m + s)^2 - \frac{1}{4} \right]a_s + a_{s-2} = 0.$$

For $s = 0$, $(m^2 - \frac{1}{4})a_0 = 0$; since $a_0 \neq 0$, we find $m = \pm\frac{1}{2}$. Then

$$\left[\left(s \pm \frac{1}{2}\right)^2 - \frac{1}{4} \right]a_s + a_{s-2} = 0$$

$$(s^2 \pm s)a_s + a_{s-2} = 0.$$

With $m = \frac{1}{2}$, we obtain $a_1 = a_3 = a_5 = \ldots = 0$, and

$$a_s = \frac{-a_{s-2}}{s(s + 1)}.$$

The solution then is

$$a_0 x^{1/2}\left[1 - \frac{x^2}{3!} + \frac{x^4}{5!} - \frac{x^6}{7!} + \cdots\right].$$

With $m = -\frac{1}{2}$, both a_0 and a_1 are arbitrary and we get, using $a_s = \dfrac{-a_{s-2}}{s(s-1)}$, the solution

$$a_0 x^{-1/2}\left[1 - \frac{x^2}{2!} + \frac{x^4}{4!} - \cdots\right] + a_1 x^{1/2}\left[1 - \frac{x^2}{3!} + \frac{x^4}{5!} - \cdots\right].$$

Clearly the general solution is

$$K\sqrt{x}\left[1 - \frac{x^2}{3!} + \frac{x^4}{5!} - \cdots\right] + \frac{L}{\sqrt{x}}\left[1 - \frac{x^2}{2!} + \frac{x^4}{4!} - \cdots\right] = K\frac{\sin x}{\sqrt{x}} + L\frac{\cos x}{\sqrt{x}}.$$

It happens that in this example the final answer can be put in a closed form; in general, this will not be the case, but the power series gives a form of the solution which, for certain values of x, is useful for computation. Thus, if we require the solution having $y = 0$ for $x = 0$, the result is

$$K\sqrt{x}\left[1 - \frac{x^2}{3!} + \frac{x^4}{5!} - \cdots\right].$$

One further condition may be satisfied.

The method of this section fails for those cases when a series solution of the form $\sum\limits_{s=0}^{\infty} a_s x^{m+s}$ does not exist; also it may fail to give the most general solution (involving n arbitrary constants). For further discussion the student is referred to H. T. H. Piaggio, *Differential Equations*, Chapter 9.

EXERCISES

1. Show that the method of this section produces a solution (but not the most general solution) of
$$x^2 y'' + xy' + (x^2 - 4)y = 0.$$

2. Show that the method does not produce any solution for the equation
$$x^4 y'' + 2x^3 y' - y = 0.$$

3. Solve the equations
   (a) $(x^2 - 4)y'' + 5xy' + y = 0$,
   (b) $(1 - x^2)y'' - 2xy' + 2y = 0$,
   (c) $x^2 y'' + xy' + (x^2 - \frac{9}{4})y = 0$,
   (d) $y' = x^2 + 4y$,
   (e) $y'' + xy = 0$.

4. Solve the Weber-Hermite equation
$$y'' - xy' + ny = 0$$
subject to the conditions $y = y' = 1$ for $x = 0$.

5. Show that, for $x > 0$, the solution of $y' - y = 1/x$ can be written as

$$y = e^x \left[A - \int_{-\infty}^{-x} \frac{e^t}{t} \, dt \right] \sim \left[A e^x - \sum_{r=1}^{\infty} (-)^r \frac{(r-1)!}{x^r} \right].$$

3. THE PICARD METHOD

The Picard method consists in taking an approximation y_1 to the solution of $dy/dx = f(x, y)$, and then obtaining a second approximation from $y_2 = \int f(x, y_1) \, dx$. The process may then be repeated as often as desired.

EXAMPLE 6.3. Tabulate, in 0(.1).5, the solution of $dy/dx = 1 + xy$ which passes through (0, 1).

Take $y_1 = 1$; then

$$y_2 = \int (1 + xy_1) \, dx = 1 + x + \frac{x^2}{2},$$

$$y_3 = \int (1 + xy_2) = 1 + x + \frac{x^2}{2} + \frac{x^3}{3} + \frac{x^4}{8},$$

$$y_4 = 1 + x + \frac{x^2}{2} + \frac{x^3}{3} + \frac{x^4}{8} + \frac{x^5}{15} + \frac{x^6}{48}.$$

Tabulating, we obtain

x	0	.1	.2	.3	.4	.5
y	1.000	1.105	1.223	1.355	1.505	1.677

Quite often we can save labour by noting that each approximation alters only one or two terms to their final form; we omit those subject to later alterations.

EXAMPLE 6.4.

$$\frac{dy}{dx} = x + y^2; \qquad (x = 0, y = 1).$$

Take

$$y_1 = 1,$$

$$y_2 = \int (x + 1) \, dx = 1 + x + \frac{x^2}{2} \doteq 1 + x,$$

$$y_3 = \int (x + y_2^2) \, dx \doteq \int (x + 1 + 2x) \, dx$$

$$\doteq 1 + x + \tfrac{3}{2}x^2,$$

$$y_4 = \int (x + y_3^2) \, dx \doteq \int (x + 1 + 2x + 4x^2) \, dx$$

$$\doteq 1 + x + \tfrac{3}{2}x^2 + \tfrac{4}{3}x^3,$$

$$y_5 = \int (x + y_4^2) \, dx \doteq \int (x + 1 + 2x + 4x^2 + \tfrac{17}{3}x^3) \, dx$$

$$\doteq 1 + x + \tfrac{3}{2}x^2 + \tfrac{4}{3}x^3 + \tfrac{17}{12}x^4.$$

Clearly there is no point in finding y^2 exactly each time, since we gain only one term in the approximation at each stage.

EXERCISES

1. Employ the Picard method to tabulate in 0(.1).5 the solution of

$$\frac{dy}{dx} = 1 + x^2 y$$

which passes through $(0, \frac{1}{2})$. Give results to four decimals.

2. Find, in 0(.1)1, the solution of

$$\frac{dy}{dx} = x + x^4 y$$

through $(0, 3)$.

3. Use the Picard method to obtain series solutions through $(0, 2)$ for

(a) $\dfrac{dy}{dx} = x^2 + y^2$,   (b) $\dfrac{dy}{dx} = x^3 - x^4 y$,

(c) $\dfrac{dy}{dx} = x^3 + y^3$.

4. SOLUTION BY REPEATED TAYLOR SERIES

The Picard method of successive integrations fails if the function is not easily integrated (although we can always employ numerical integration, the necessity for doing so at each integration is unattractive). However, if we need a solution through (x_0, y_0), the Taylor series at once tells us that the answer is

$$y = y_0 + y_0' \frac{(x - x_0)}{1!} + y_0'' \frac{(x - x_0)^2}{2!} + y_0''' \frac{(x - x_0)^3}{3!} + \dots,$$

and the derivatives y_0', y_0'', ... can always be calculated from the given equation (6.1).

EXAMPLE 6.5.

$$5xy' + y^2 - 2 = 0; \qquad x_0 = 4, \quad y_0 = 1.$$

Differentiating:

$$5xy'' + 5y' + 2yy' = 0,$$
$$5xy''' + 10y'' + 2yy'' + 2y'^2 = 0,$$
$$5xy^{iv} + 15y''' + 2yy''' + 6y'y'' = 0,$$
$$5xy^{v} + 20y^{iv} + 2yy^{iv} + 8y'y''' + 6y''^2 = 0.$$

Solving from these equations in succession:

$$x_0 = 4, \quad y_0 = 1, \quad y_0' = .05, \quad y_0'' = -.0175,$$

$$y_0''' = .01025, \quad y_0^{iv} = -.00845, \quad y_0^v = .008998125.$$

Then

$$y = 1 + .05(x - 4) - .00875(x - 4)^2 + .0017083(x - 4)^3$$
$$- .0003521(x - 4)^4 + .00007498(x - 4)^5 + \ldots$$

Tabulating in 4(.1)4.4, we get

x	y
4	1.000000
4.1	1.004914
4.2	1.009663
4.3	1.014256
4.4	1.018701

We have kept six decimals with a view to using this example again in the next section.

After we reach a certain distance from the initial value, convergence of the Taylor series slows up; we can then repeat the process to get a second series. If we now take $x_0 = 4.4$, $y_0 = 1.018701$, we get

$$y_0 = 1.018701, \quad y_0' = .043739, \quad y_0'' = -.01399, \quad y_0''' = .00748.$$

Then

$$y = 1.018701 + .043739(x - 4.4) - .00700(x - 4.4)^2$$
$$+ .00125(x - 4.4)^3 + \ldots$$

For $x = 4.5$, we get

$$y = 1.018701 + .004374 - .000070 + .000001 = 1.023006.$$

Note that the work is relatively light if we use short cuts such as $1.018701^2 = (1 + .018701)^2 = 1 + 2(.018701) + .0187^2$. A table of squares now removes all labour.

EXERCISES

1. Take Example 6.1,

$$10xy' + y^2 - 2 = 0; \quad (x = 4, y = 1),$$

and tabulate the solution in 4(.2)5 to four decimals.

2. Find the Taylor series for Example 6.1 about the point (5, 1.0218), and use it to continue the solution as far as 5.4.

3. $10x^2y' + y^2 - 4 = 0$; $(5, 1)$. Find the Taylor series solution about $x = 5$.

4. Solve the following equations by obtaining the Taylor series solution.

(a) $\dfrac{dy}{dx} = x^2 + y^2$; $(0, \tfrac{1}{2})$,   (b) $\dfrac{dy}{dx} = \dfrac{1}{x^2 + y}$; $(4, 4)$.

5. The equation $dy/dx = x^2 + y$ is to be solved, subject to the initial condition that the solution pass through the point $(0, 10)$. Solve this equation by expressing it as $\dfrac{dx}{dy} = \dfrac{1}{x^2 + y}$ and obtaining x as a Taylor series in y. This switch is often efficacious in cases where y_0' is large, as in this exercise.

5. THE ADAMS-BASHFORTH PROCESS

The Adams process is based on a combination of the Taylor series approach of Section 4 with the use of finite differences. We have, for an arbitrary function q,

$$q_{n+kh} = E^k q_n = \left(\frac{1}{E}\right)^{-k} q_n = \left(\frac{E - \Delta}{E}\right)^{-k} q_n = (1 - \Delta E^{-1})^{-k} q_n.$$

Expanding gives (h the interval of differencing)

(6.4)

$$q_{n+kh} = q_n + k\,\Delta q_{n-1} + \frac{k(k + 1)}{2!}\Delta^2 q_{n-2} + \frac{k(k + 1)(k + 2)}{3!}\Delta^3 q_{n-3} + \cdots$$

(Of course, Formula (6.4) can be obtained slightly more readily by using the backward difference operator ∇.) Now set $q = h\,dy/dx$, and integrate from $k = 0$ to 1. We get

$$y_{n+1} - y_n = \int_0^1 y'_{n+kh}(h\,dk)$$

$$= \int_0^1 q_{n+kh}\,dk$$

$$= \int_0^1 \left[q_n + k\,\Delta q_{n-1} + \frac{k(k + 1)}{2!}\Delta^2 q_{n-2} + \cdots\right] dk.$$

Thus

(6.5) $y_{n+1} - y_n = q_n + \tfrac{1}{2}\Delta q_{n-1} + \tfrac{5}{12}\Delta^2 q_{n-2} + \tfrac{3}{8}\Delta^3 q_{n-3} + \cdots$

Formula (6.5) is the basic formula in the Adams method; it allows one to obtain y_{n+1} when given y_n and the differences of the function hy'. These last should be small.

EXAMPLE 6.6.

$$5xy' + y^2 - 2 = 0; \qquad (4, 1).$$

We need some initial values, and obtain them by the Taylor series method used in Example 6.5 of the last section. The function q is computed from

$$q = hy' = h\frac{2 - y^2}{5x} = \frac{2 - y^2}{50x}.$$

The previous remark about computing y^2 by the binomial theorem is again relevant.

$10^6 y$	$10^7 q$	$10^7 \Delta q$	$10^7 \Delta^2 q$	$10^7 \Delta^3 q$
1000000				
	50000			
1004914		−1700		
	48300		94	
1009663		−1606		−6
	46694		88	
1014256		−1518		−7
	45176		81	
1018701		−1437		−7
	43739		74	
1023006		−1363		
	42376			
1027178				

Using the part of the table above the solid line, we get

$$y_5 = y_4 + q_4 + \tfrac{1}{2}\Delta q_3 + \tfrac{5}{12}\Delta^2 q_2 + \tfrac{3}{8}\Delta^3 q_1 + \cdots$$
$$= 1.018701 + .0043739 - .0000719 + .0000034 - .0000004$$
$$= 1.023006.$$

We then compute $q_5 = .0042376$, and fill in the line of the table below the solid line. Then

$$y_6 = y_5 + q_5 + \tfrac{1}{2}\Delta q_4 + \tfrac{5}{12}\Delta^2 q_3 + \cdots$$
$$= 1.027178.$$

The process is now continued line by line.

EXERCISES

1. Continue the solution of Example 6.6 to obtain y for $x = 5$.

2. Take the equation
$$10xy' + y^2 - 2 = 0, \quad (4, 1),$$
given in Exercise 1 of the last section, and continue the solution from 5 to 6.

3. Obtain the term in $\Delta^4 q_{n-4}$ in Formula (6.5).

6. THE MILNE METHOD

The Milne method is based on the open and closed Simpson rules applied to the function q. Thus we have

$$\int_0^2 h y'_{kh} \, dk = \frac{1}{3}(q_0 + 4q_1 + q_2),$$

that is,

$$y_2 = y_0 + \frac{1}{3}(q_0 + 4q_1 + q_2);$$

more generally,

(6.6) $$y_{n+1} = y_{n-1} + \frac{1}{3}(q_{n-1} + 4q_n + q_{n+1}).$$

The open Simpson rule gives us

$$\int_0^4 h y'_{kh} \, dk = \frac{4}{3}(2q_1 - q_2 + 2q_3),$$

that is,

$$y_4 - y_0 = \frac{4}{3}(2q_1 - q_2 + 2q_3);$$

more generally,

(6.7) $$y_{n+1} = y_{n-3} + \frac{4}{3}(2q_n - q_{n-1} + 2q_{n-2}).$$

The method involves using (6.7) as a "predictor" to give y_{n+1} from known q's and y's. From this predicted value and the differential equation, q_{n+1} is computed; knowing q_{n+1}, (6.6) can then be used as a "corrector" to give a better value of y_{n+1}. Then q_{n+1} can be recalculated and (6.6) again applied. The process of using (6.7) and (6.6) alternately is repeated until the table of values is complete.

EXAMPLE 6.7. We again use the illustration

$$5xy' + y^2 - 2 = 0; \quad (4, 1).$$

The table of y's and q's is reproduced from the last section.

$10^6 y$	$10^7 q$
1000000	50000
1004914	48300
1009663	46694
1014256	45176
1018701	43739

Using (6.7) as predictor,

$$y_5 = y_1 + \tfrac{4}{3}(2q_4 - q_3 + 2q_2)$$
$$= 1.004914 + .018092 = 1.023006.$$

Then

$$q_5 = \frac{2 - y^2}{50x} = .0042376.$$

Using (6.6) as corrector, we obtain

$$y_5 = y_3 + \tfrac{1}{3}(q_3 + 4q_4 + q_5)$$
$$= 1.014256 + .008750 = 1.023006.$$

Since the corrector has given the same result as the predictor, we can accept the value. Had the corrected y_5 been different, q_5 would have been recomputed and (6.6) reapplied.

EXERCISES

1. Use the Milne method to continue Example 6.7 as far as the y-value corresponding to $x = 5$.

2. Work Exercise 2 of Section 5 by the Milne method.

7. THE RUNGE-KUTTA METHOD

There are vast numbers of methods for numerical solutions of differential equations; the student will find exceptionally lucid accounts of many methods given in H. Levy and E. A. Baggott, *Numerical Solutions of Differential Equations.* To conclude this chapter, we shall describe the method of Runge and Kutta, which has considerable historical importance in that it was one of the earliest methods developed.

We have seen that the Taylor expansion of a function y can be written

$$\Delta y = (e^{hD} - 1)y$$
$$= \left(hD + \frac{h^2 D^2}{2!} + \frac{h^3 D^3}{3!} + \frac{h^4 D^4}{4!} + \dots \right) y.$$

Apply this to the differential equation (6.1),

$$\frac{dy}{dx} = f(x, y),$$

and obtain

$$\Delta y = hf + \frac{h^2}{2} Df + \frac{h^3}{6} D^2 f + \frac{h^4}{24} D^3 f + \dots,$$

where we write f for $f(x, y)$. Using subscripts 1 and 2 to denote partial differentiation with respect to x and y respectively, this last equation may be written

$$(6.8) \quad \Delta y = hf + \frac{h^2}{2}(f_1 + ff_2) + \frac{h^3}{6}[f_{11} + 2ff_{12} + f^2f_{22} + f_2(f_1 + ff_2)]$$

$$+ \frac{h^4}{24}[f_{111} + 3ff_{112} + 3f^2f_{122} + f^3f_{222} + f_2(f_{11} + 2ff_{12} + f^2f_{22})$$

$$+ 3(f_1 + ff_2)(f_{12} + ff_{22}) + f_2^2(f_1 + ff_2)] + \cdots$$

A Runge-Kutta formula of order n is a formula which expresses Δy in terms of n values of the function $f(x, y)$ in such a manner that the value obtained coincides with Equation (6.8) as far as the terms involving h^n. We shall develop a fourth-order formula; to that end, we define

$$(6.9) \quad \begin{cases} k_1 = hf(x, y) = hf, \\ k_2 = hf(x + mh, y + mk_1), \\ k_3 = hf(x + nh, y + nk_2), \\ k_4 = hf(x + ph, y + pk_3). \end{cases}$$

Our aim will then be to express Δy in the form

$$(6.10) \quad \Delta y = ak_1 + bk_2 + ck_3 + dk_4.$$

First, we introduce some useful abbreviations; let us write

$$F_1 = f_1 + ff_2,$$

$$F_2 = f_{11} + 2ff_{12} + f^2f_{22},$$

$$F_3 = f_{111} + 3ff_{112} + 3f^2f_{122} + f^3f_{222};$$

then (6.8) can be abbreviated to read

$$(6.11) \quad \Delta y = hf + \frac{h^2}{2}F_1 + \frac{h^3}{6}(F_2 + f_2F_1)$$

$$+ \frac{h^4}{24}(F_3 + f_2F_2 + 3f_2'F_1 + f_2^2F_1) + \cdots$$

Furthermore, we may use the Taylor series for two variables to compute

$$k_1 = hf,$$

$$k_2 = h\left[f + mhF_1 + \frac{m^2h^2}{2}F_2 + \frac{m^3h^3}{6}F_3 + \ldots\right],$$

$$k_3 = h\left[f + nhF_1 + \frac{h^2}{2}(n^2F_2 + 2mnf_2F_1)\right.$$

$$\left. + \frac{h^3}{6}(n^3F_3 + 3m^2nf_2F_2 + 6mn^2f_2'F_1) + \ldots\right],$$

$$k_4 = h\left[f + phF_1 + \frac{h^2}{2}(p^2F_2 + 2npf_2F_1)\right.$$

$$\left. + \frac{h^3}{6}(p^3F_3 + 3n^2pf_2F_2 + 6np^2f_2'F_1 + 6mnpf_2^2F_1) + \ldots\right].$$

Applying (6.10), we compute Δy and compare with (6.11); equating coefficients of corresponding expressions yields the equations

$$
\begin{aligned}
a + b + c + d &= 1, & cmn + dnp &= \tfrac{1}{6}, \\
bm + cn + dp &= \tfrac{1}{2}, & cmn^2 + dnp^2 &= \tfrac{1}{8}, \\
bm^2 + cn^2 + dp^2 &= \tfrac{1}{3}, & cm^2n + dn^2p &= \tfrac{1}{12}, \\
bm^3 + cn^3 + dp^3 &= \tfrac{1}{4}, & dmnp &= \tfrac{1}{24}.
\end{aligned}
$$

(6.12)

Any solution of Equations (6.12) will work; suppose we take $m = n = \tfrac{1}{2}$, $p = 1$, with $a = d = \tfrac{1}{6}$, $b = c = \tfrac{1}{3}$. Then (6.12) is satisfied, and

$$k_1 = hf(x, y),$$

$$k_2 = hf\left(x + \frac{h}{2}, y + \frac{k_1}{2}\right),$$

$$k_3 = hf\left(x + \frac{h}{2}, y + \frac{k_2}{2}\right),$$

$$k_4 = hf(x + h, y + k_3).$$

Thus

(6.13) $$\Delta y = \tfrac{1}{6}(k_1 + 2k_2 + 2k_3 + k_4),$$

that is, (6.13) expresses Δy in terms of the four ordinates k_i ($i = 1, 2, 3, 4$), and the expression agrees with the exact expression (6.11) up to and including the terms in h^4.

EXAMPLE 6.8.

$$\frac{dy}{dx} = 3x + \tfrac{1}{2}y; \quad (0, 1).$$

We note that dy/dx is initially less than one; had it been large, we might have preferred to work with dx/dy in order to deal with a function which changed at a more moderate pace.

We take $h = .1$; then

$$k_1 = hf(0, 1) = .05,$$
$$k_2 = hf(.05, 1.025) = .06625,$$
$$k_3 = hf(.05, 1.033125) = .06665625,$$
$$k_4 = hf(.1, 1.06665625) = .0833328125.$$

Consequently,

$$\Delta y = \tfrac{1}{6}(k_1 + 2k_2 + 2k_3 + k_4) = .06652421875,$$

and

$$y(.1) = 1.06652421875.$$

The process may now be continued, starting from this value.

We can gain an insight into the accuracy of the Runge-Kutta formula (6.13) by continuing Example 6.8 to give the exact solution. It is easy to find that this exact solution is

$$y = 13e^{x/2} - 6x - 12,$$

and

$$y(.1) = 13e^{.05} - 12.6 = 1.06652425289.$$

There is a difference of about three units in the eighth decimal place. On the other hand, if we take $h = .2$, the Runge-Kutta Formula $y(.2) = 1.167220833$, as compared to the correct value 1.167221935; whereas taking $h = .4$ yields $y(.4) = 1.4782$ as opposed to the true value 1.478234. These results show how decreasing h diminishes the error in the Runge-Kutta Formula.

The question of determining the accuracy of the results of numerical solution of a differential equation is quite difficult; the beginner should use a small value for h, since decreasing h greatly improves the accuracy. One very good check is to work the problem with a value h, and then repeat using $h/2$; this is the same procedure recommended in the discussion of Simpson's Rule, and it is very efficacious.

EXERCISES

1. Show that Formula (6.13) reduces to Simpson's Rule if the function $f(x, y)$ does not involve y at all.

2. Tabulate the exact solution of Example 6.8 to five decimals in $0(.1).4$. Compare the values given by the Runge-Kutta formula (6.13).

3. Obtain a Runge-Kutta formula by putting $m = \tfrac{1}{3}$, $n = \tfrac{2}{3}$, $p = 1$, in (6.12); what does this formula reduce to in the case that $f(x, y)$ is independent of y?

4. Obtain a general third-order Runge-Kutta formula by putting

$$k_1 = hf(x, y),$$
$$k_2 = hf(x + mh, y + mk_1),$$
$$k_3 = hf(x + nh, y + pk_1 + qk_2).$$

5. Use the Runge-Kutta formula (6.13) to solve the equation

$$10 \frac{dy}{dx} = x^2 + y^2, \quad (0, 1),$$

in the interval $0(.1).8$.

6. Solve the equation

$$8 \frac{dy}{dx} = x + y^2, \quad (0, .5),$$

in the interval $(0, 1)$ by taking h successively as $.2, .1, .05$.

7. Develop a slightly more general fourth-order Runge-Kutta formula by taking

$$k_1 = h f(x, y),$$
$$k_2 = h f(x + mh, y + mk_1),$$
$$k_3 = h f(x + nh, y + rk_2 + [r - n]k_1),$$
$$k_4 = h f(x + ph, y + qk_3 + sk_2 + [p - q - s]k_1).$$

Show that

$$k_2 = h \left[f + mhF_1 + \frac{m^2h^2}{2} F_2 + \frac{m^3h^3}{6} F_3 + \ldots \right],$$

$$k_3 = h \left[f + nhF_1 + \frac{h^2}{2} (n^2F_2 + 2mrf_2F_1) \right.$$
$$\left. + \frac{h^3}{6} (n^2F_3 + 3m^2rf_2F_2 + 6mnrf'_2F_1) + \ldots \right],$$

$$k_4 = h \left[f + phF_1 + \frac{h^2}{2} (p^2F_2 + 2f_2\{qn + sm\}F_1) \right.$$
$$+ \frac{h^3}{6}(p^3F_3 + 3f_2\{qn^2 + sm^2\}F_2 + 6f_2^2\, rmqf_2F_1$$
$$\left. + 6p\{qn + sm\}f'_2F_1) + \ldots \right].$$

Deduce that the constants in the formula satisfy the equations

$$a + b + c + d = 1, \qquad crm + d(qn + sm) = \tfrac{1}{6},$$
$$bm + cn + dp = \tfrac{1}{2}, \qquad crmn + pd(qn + sm) = \tfrac{1}{8},$$
$$bm^2 + cn^2 + dp^2 = \tfrac{1}{3}, \qquad crm^2 + d(qn^2 + sm^2) = \tfrac{1}{12},$$
$$bm^3 + cn^3 + dp^3 = \tfrac{1}{4}, \qquad dmqr = \tfrac{1}{24}.$$

8. Apply Formula (6.13) to solve

$$\frac{dy}{dx} = \frac{x + y}{10e^x}, \quad (0, 5),$$

in the interval $(0, 1)$ by taking $h = .2$ and then repeating with $h = .1$. Repeat, using the Adams-Bashforth process.

CHAPTER SEVEN

Linear Systems and Matrices

1. REVIEW OF DETERMINANTS

In this section, we shall recall (without proofs) the various properties of determinants.

In appearance, a *determinant of order n* is an array of n^2 numbers arranged in n rows and n columns, and enclosed by two vertical bars. Thus

$$\begin{vmatrix} -2 & 0 & 2 \\ 3 & 1 & 7 \\ 4 & 6 & 5 \end{vmatrix}$$

is a determinant of order 3. We usually write the general determinant of order n in the form

(7.1)
$$D = \begin{vmatrix} a_{11} & a_{12} & a_{13} & \cdots & a_{1n} \\ a_{21} & a_{22} & a_{23} & \cdots & a_{2n} \\ a_{31} & a_{32} & a_{33} & \cdots & a_{3n} \\ \cdot & \cdot & \cdot & \cdot & \cdot \\ a_{n1} & a_{n2} & a_{n3} & \cdots & a_{nn} \end{vmatrix} .$$

The typical element in D is a_{ij}, where the first subscript refers to the row and the second to the column; thus a_{ij} means the element in row i and column j of D. Definition (7.1) is sometimes written in the briefer form

(7.2) $$D = |a_{ij}|_{n \times n},$$

where we write only a typical element and the dimensions of the determinant.

To every element a_{ij} in D corresponds its *cofactor* A_{ij} defined by

(7.3) $$A_{ij} = (-)^{i+j} M_{ij},$$

where M_{ij} is the determinant (or *minor*) left after deleting row i and column j from D. For example, in

$$\begin{vmatrix} 3 & -2 & 6 \\ 0 & 1 & 4 \\ 7 & 8 & 5 \end{vmatrix},$$

$$a_{23} = 4, \qquad A_{23} = (-)^{2+3} \begin{vmatrix} 3 & -2 \\ 7 & 8 \end{vmatrix};$$

$$a_{31} = 7, \qquad A_{31} = (-)^{3+1} \begin{vmatrix} -2 & 6 \\ 1 & 4 \end{vmatrix}.$$

We are now in a position to define the numerical value of the determinant (7.1).

1) If $D = |a|$, then $D = a$. (This assigns a numerical value to a determinant of order 1.)

2) If D is a determinant of order n, then its value is found by selecting any line (row or column), and taking the sum of all elements in that line, each multiplied by its cofactor. In symbols,

(7.4) $$D = \sum_{j=1}^{n} a_{ij} A_{ij} \quad \text{(summing along row } i\text{)},$$

or

(7.5) $$D = \sum_{i=1}^{n} a_{ij} A_{ij} \quad \text{(summing along column } j\text{)}.$$

Definitions (7.4) and (7.5) allow us to evaluate any determinant; for example, a determinant of order five can be expressed in terms of five determinants of order four; each determinant of order four can be expressed in terms of four determinants of order three, etc. Ultimately, the determinant of order five is expressed in terms of $5! = 60$ determinants of order one, and the values of these are known.

If we were to base a rigorous development of determinant theory on (7.4) and (7.5), we should need to include a proof of the

Uniqueness Theorem: The numerical value obtained for D is independent of the particular row or column used in (7.4) or (7.5).

Such a proof can be given by mathematical induction, but, since it is quite long, we omit it here.

EXAMPLE 7.1. Let

$$D = \begin{vmatrix} a_{11} & a_{12} \\ a_{21} & a_{22} \end{vmatrix}.$$

Using the first row,

$$D = a_{11}A_{11} + a_{12}A_{12}$$
$$= a_{11}(-)^{1+1}a_{22} + a_{12}(-)^{1+2}a_{21}$$
$$= a_{11}a_{22} - a_{12}a_{21}.$$

This result gives us the rule: the value of a determinant of *order two* is found by taking the downward cross-product $a_{11}\,a_{22}$ less the upward cross-product $a_{12}\,a_{21}$. Thus

$$\begin{vmatrix} 2 & 3 \\ -6 & 8 \end{vmatrix} = 16 - (-18) = 34.$$

The student should avoid any attempt to use this rule for other than determinants of order two.

EXAMPLE 7.2.

$$\begin{vmatrix} 3 & -2 & 4 \\ 1 & 7 & 8 \\ 2 & -9 & 3 \end{vmatrix}$$

Using the second row,

$$D = 1(-)^3\begin{vmatrix} -2 & 4 \\ -9 & 3 \end{vmatrix} + 7(-)^4\begin{vmatrix} 3 & 4 \\ 2 & 3 \end{vmatrix} + 8(-)^5\begin{vmatrix} 3 & -2 \\ 2 & -9 \end{vmatrix}$$
$$= -30 + 7 + 184 = 161.$$

Using the third column (as a check),

$$D = 4(-)^4\begin{vmatrix} 1 & 7 \\ 2 & -9 \end{vmatrix} + 8(-)^5\begin{vmatrix} 3 & -2 \\ 2 & -9 \end{vmatrix} + 3(-)^6\begin{vmatrix} 3 & -2 \\ 1 & 7 \end{vmatrix}$$
$$= -92 + 184 + 69 = 161.$$

Clearly it would be clumsy and time-consuming to use cofactors every time we wanted to work out a determinant; the process of reducing the order of the determinant by repeated use of Equations (7.4) and (7.5) is direct but slow. We now state and exemplify some useful properties of determinants which serve as short cuts. Again we omit proofs, which can be based on Definitions (7.4) and (7.5).

LEMMA 7.1. If we interchange two parallel lines in a determinant D, then the value of the new determinant is $-D$.

EXAMPLE 7.3.

$$D = \begin{vmatrix} 1 & 3 & 2 \\ 0 & 5 & -6 \\ 2 & 7 & 8 \end{vmatrix} = 1(-)^2 \begin{vmatrix} 5 & -6 \\ 7 & 8 \end{vmatrix} + 3(-)^3 \begin{vmatrix} 0 & -6 \\ 2 & 8 \end{vmatrix} + 2(-)^4 \begin{vmatrix} 0 & 5 \\ 2 & 7 \end{vmatrix}$$

$$= 82 - 36 - 20 = 26.$$

$$D_1 = \begin{vmatrix} 2 & 7 & 8 \\ 0 & 5 & -6 \\ 1 & 3 & 2 \end{vmatrix} = 0 + 5(-)^4 \begin{vmatrix} 2 & 8 \\ 1 & 2 \end{vmatrix} - 6(-)^5 \begin{vmatrix} 2 & 7 \\ 1 & 3 \end{vmatrix}$$

$$= -20 - 6 = -26.$$

LEMMA 7.2. If we add λ times any line of a determinant D to any parallel line, then the new determinant has the same value as D.

EXAMPLE 7.4.

$$D = \begin{vmatrix} 3 & -6 & 5 \\ 0 & 2 & 1 \\ 7 & -4 & 8 \end{vmatrix} = 0 + 2 \begin{vmatrix} 3 & 5 \\ 7 & 8 \end{vmatrix} - 1 \begin{vmatrix} 3 & -6 \\ 7 & -4 \end{vmatrix} = -22 - 30 = -52.$$

Form E by adding three times the second row to the first row. Then

$$E = \begin{vmatrix} 3 & 0 & 8 \\ 0 & 2 & 1 \\ 7 & -4 & 8 \end{vmatrix} = 3 \begin{vmatrix} 2 & 1 \\ -4 & 8 \end{vmatrix} + 8 \begin{vmatrix} 0 & 2 \\ 7 & -4 \end{vmatrix} = -52 = D.$$

Less important rules, which follow directly from Lemmas 7.1 and 7.2, or from Formulae (7.4) and (7.5), are

LEMMA 7.3. If a determinant E is formed from D by multiplying every element of some line of D by a constant λ, then $E = \lambda D$.

COROLLARY. If D contains a line of zeros, then $D = 0$.

LEMMA 7.4. If we define the transpose of D to·be the determinant D' whose ith column is the ith row of D ($i = 1, 2, \ldots , n$), then $D' = D$.

LEMMA 7.5. If two parallel lines of D are identical, then $D = 0$.

LEMMA 7.6. If we add up the elements of any line, each multiplied by the cofactors of a parallel line, the result is zero. In symbols,

$$\sum_{j=1}^{n} a_{ij}A_{kj} = \sum_{i=1}^{n} a_{ij}A_{ik} = 0.$$

If a determinant is complicated numerically, it is always very difficult to evaluate. In simple cases, the most efficacious method is to employ Lemma 7.2 alternately with Equations (7.4) and (7.5).

EXAMPLE 7.5.

$$D = \begin{vmatrix} 3 & 2 & 4 & 0 \\ -2 & 1 & 3 & 5 \\ 8 & 7 & -2 & 4 \\ 5 & -3 & 3 & 6 \end{vmatrix}.$$

Add -2, -7, and 3 times the second row to the first, third, and fourth rows respectively, and then expand by cofactors of the second column.

$$\begin{vmatrix} 7 & 0 & -2 & -10 \\ -2 & 1 & 3 & 5 \\ 22 & 0 & -23 & -31 \\ -1 & 0 & 12 & 21 \end{vmatrix} = \begin{vmatrix} 7 & -2 & -10 \\ 22 & -23 & -31 \\ -1 & 12 & 21 \end{vmatrix}.$$

Add 7 and 22 times the third row to the first and second respectively. Then

$$D = \begin{vmatrix} 0 & 82 & 137 \\ 0 & 241 & 431 \\ -1 & 12 & 21 \end{vmatrix} = - \begin{vmatrix} 82 & 137 \\ 241 & 431 \end{vmatrix} = -2325.$$

EXERCISES

1.
$$D = \begin{vmatrix} 3 & 0 & -1 & 0 \\ 2 & 7 & 5 & 4 \\ 0 & 5 & 8 & 2 \\ 1 & 6 & 7 & 0 \end{vmatrix}$$

(a) Evaluate A_{11} and A_{13}, and deduce the value of D.
(b) Evaluate A_{24} and A_{34}, and deduce the value of D.

2. Evaluate
$$\begin{vmatrix} 3 & 6 & -2 & 0 \\ 8 & 0 & 1 & 9 \\ 5 & 1 & 2 & -12 \\ -3 & 8 & 1 & -4 \end{vmatrix}.$$

3. Prove Lemmas 7.3 and 7.4 directly from Formulae (7.4) and (7.5).

4. (a) Prove Lemma 7.5 using Lemma 7.2 and Formula (7.4).
   (b) Prove Lemma 7.5 using Lemma 7.1.

5. Prove Lemma 7.6 using Lemma 7.5 and Formula (7.4).
   Hint: Form a determinant E from D such that E is identical with D except that the kth row of E contains the elements from the ith row of D.

6. Show that if n is odd and $a_{ij} + a_{ji} = 0$, then $D = 0$.

7. If a, b, and c are sides of a triangle, evaluate

$$\begin{vmatrix} x & y & z \\ a & b & c \\ \sin A & \sin B & \sin C \end{vmatrix}.$$

8. Evaluate

$$\begin{vmatrix} 1 & 2 & 0 & 4 \\ 3 & 0 & -1 & -2 \\ 0 & -1 & 3 & 0 \\ 5 & 0 & 1 & 1 \end{vmatrix}.$$

9. A determinant has $a_{ij} = B$ $(i \neq j)$ and $a_{ii} = A$. Prove

$$D = [A + (n-1)B][A - B]^{n-1}.$$

2. CRAMER'S SOLUTION OF n LINEAR EQUATIONS

Suppose we have n linear equations in n unknowns of the form

(7.6) $$\sum_{j=1}^{n} a_{ij}x_j = c_i \qquad (i = 1, 2, \dots, n).$$

Now define determinants D and D_k as follows:

$D = |a_{ij}|_{n \times n}$; D_k is exactly the same as D except that the kth column of D has been replaced by the column of constants $c_1, c_2, \dots, c_n$ to give D_k.

Then we obtain

THEOREM 7.1 (Cramer). Equations (7.6) are solved by

$$x_k = \frac{D_k}{D}.$$

Proof. Multiply (7.6) by A_{ik} and add over i.

$$\sum_{i=1}^{n} A_{ik} \sum_{j=1}^{n} a_{ij}x_j = \sum_{i=1}^{n} c_i A_{ik}.$$

Reverse the order of summation.

$$\sum_{j=1}^{n} x_j \sum_{i=1}^{n} a_{ij} A_{ik} = \sum_{i=1}^{n} c_i A_{ik}.$$

Using (7.5) and Lemma 7.6, we have

$$\sum_{i=1}^{n} c_i A_{ik} = D_k, \qquad \sum_{i=1}^{n} a_{ij} A_{ik} = D\delta_{jk},$$

where δ_{jk} is the Kronecker Delta, $\delta_{jk} = 1$ $(j = k)$, $\delta_{jk} = 0$ $(j \neq k)$, Then

$$\sum_{j=1}^{n} x_j D\delta_{jk} = D_k,$$

$$x_k D\delta_{kk} = D_k,$$

that is, $Dx_k = D_k$. Consequently, $x_k = D_k/D$, as required.

EXAMPLE 7.6.

$$3x - 4y + 5z = 8,$$
$$x + 2y - 6z = 7,$$
$$2x - y + 5z = 3.$$

Here

$$D = \begin{vmatrix} 3 & -4 & 5 \\ 1 & 2 & -6 \\ 2 & -1 & 5 \end{vmatrix} = \begin{vmatrix} 0 & -10 & 23 \\ 1 & 2 & -6 \\ 0 & -5 & 17 \end{vmatrix} = 55,$$

$$D_1 = \begin{vmatrix} 8 & -4 & 5 \\ 7 & 2 & -6 \\ 3 & -1 & 5 \end{vmatrix} = \begin{vmatrix} -4 & 0 & -15 \\ 13 & 0 & 4 \\ 3 & -1 & 5 \end{vmatrix} = 179,$$

$$D_2 = \begin{vmatrix} 3 & 8 & 5 \\ 1 & 7 & -6 \\ 2 & 3 & 5 \end{vmatrix} = \begin{vmatrix} 0 & -13 & 23 \\ 1 & 7 & -6 \\ 0 & -11 & 17 \end{vmatrix} = -32,$$

$$D_3 = \begin{vmatrix} 3 & -4 & 8 \\ 1 & 2 & 7 \\ 2 & -1 & 3 \end{vmatrix} = \begin{vmatrix} 0 & -10 & -13 \\ 1 & 2 & 7 \\ 0 & -5 & -11 \end{vmatrix} = -45.$$

Then

$$x = \frac{179}{55} = 3.255, \quad y = \frac{-32}{55} = -.545, \quad z = \frac{-45}{55} = -.818.$$

We conclude this section with a few remarks concerning the role of determinants in mathematics; it is clear that, unless the entries are small integers, determinants are cumbersome in numerical work. However, they play a useful part in mathematical theory; they are a very useful abbreviation in statistics, algebra, and geometry. It is often handy to be able to write down the *formal* solution of a set of equations as $x_k = D_k/D$; this expression, however, is very ill-suited for numerical computation. (An exception might be made for the case of two equations in two unknowns; here, one might find determinants useful as a mnemonic.)

Some engineers and pure mathematicians, not closely associated with practical numerical work, labour under the illusion that determinants are a practical method of solving equations. This belief can not be supported. A count of the number of multiplications involved in solving a system of n equations by evaluating the $n + 1$ determinants required, as compared with the much smaller number of multiplications involved in solving the system of equations by an eliminative method, speedily establishes the inferiority of the determinantal approach. Even electronic computers, which devote much of their time to solving linear systems of the form (7.6), find determinants too awkward to employ. The student should regard determinants as useful tools in developing theory; in practical numerical analysis, he should disregard them.

It might be well to stress the viewpoint of the last two paragraphs by including two quotations.

"The exact solution of the equations (7.6) is $x_k = D_k/D$. Although this is formally the complete answer, it is completely useless for practical computation." (A. D. Booth, *Numerical Methods*)

"Though there may be no one best way of evaluating the solution of the equations (7.6), it can be said with some certainty that the direct evaluation of the determinants and of the expression for the solution in terms of them is *never* the best way." (D. R. Hartree, *Numerical Analysis*)

EXERCISES

1. If
$$x' = x \cos \theta - y \sin \theta,$$
$$y' = x \sin \theta + y \cos \theta,$$
express x and y in terms of x' and y'.

2. Solve by determinants
$$2x - y + 5z = 6,$$
$$3x + 4y - 8z = 17,$$
$$4x + 2y - 7z = -8.$$

3. Show that if Equations (7.6) are *homogeneous*, that is, if $c_i = 0$ for all i, and if there exists a non-trivial solution (that is, a solution other than $x_1 = x_2 = \ldots = x_n = 0$), then $D = |a_{ij}| = 0$.

4. Find k so that the equations

$$kx - y + 2 = 0,$$
$$(k + 1)x + 6y - 5 = 0,$$
$$(k + 2)x + 3y + 4 = 0,$$

shall be consistent. (Consider them as three homogeneous equations for x, y, 1.)

5. Solve the equations

$$3x + 2y - z + u = 0,$$
$$2x + 5y + 7z - 2u = 0,$$
$$4x - y - 9z + 4u = 0,$$
$$x + 8y + 15z - 5u = 0.$$

3. MATRIC NOTATION

Suppose that m quantities $y_1, y_2, \ldots, y_m$ are related to n quantities $x_1, x_2, \ldots, x_n$ by

$$(7.7) \qquad y_i = \sum_{j=1}^{n} a_{ij} x_j \quad (i = 1, 2, \ldots, m).$$

We agree to consider the m quantities y_i as a single entity called a *column vector*

$$(7.8) \qquad Y = \begin{pmatrix} y_1 \\ y_2 \\ \cdot \\ \cdot \\ \cdot \\ y_m \end{pmatrix}.$$

The y_i are the *coordinates* or *components* of the vector. Similarly, we use X to represent the column vector with components $x_1, x_2, \ldots, x_n$. The array of coefficients

$$A = (a_{ij}) = \begin{pmatrix} a_{11} & a_{12} & \cdots & a_{1n} \\ a_{21} & a_{22} & \cdots & a_{2n} \\ \cdot & \cdot & \cdots & \cdot \\ a_{m1} & a_{m2} & \cdots & a_{mn} \end{pmatrix}$$

is called a *matrix*, and the linear transformation (7.7) is usually written in the condensed but completely equivalent form

$$(7.9) \qquad Y = AX.$$

It must be stressed that (7.9) is just a symbolic way of writing (7.7); *the i'th component of Y is found by applying the i'th row of A to the column vector X and summing all the products thus obtained.*

The notation (7.9) has, in addition to the advantage of brevity, the virtue of stressing the fact that *the linear transformation* (7.7) *depends only on the matrix A of coefficients*, not on the particular letters x_i and y_i which we may happen to use to denote the coordinates of X and Y (sometimes called the object and image vectors). Thus, the matrix A is essentially an abstract representation of the linear transformation (7.7). Since A merely represents the linear transformation (7.7), it is natural to make the

DEFINITION. Two matrices A and B (of the same dimensions) are equal if and only if all corresponding entries are equal, that is,

$$a_{ij} = b_{ij} \quad (i = 1, 2, \ldots, m; j = 1, 2, \ldots, n).$$

EXAMPLE 7.7. The matrix

$$A = \begin{pmatrix} 2 & 3 & -1 \\ 4 & 2 & 7 \end{pmatrix}$$

transforms the object vector

$$\begin{pmatrix} 3 \\ 1 \\ 3 \end{pmatrix}$$

into the image vector

$$\begin{pmatrix} 6 \\ 35 \end{pmatrix}.$$

In expanded form, A represents the linear transformation

$$y_1 = 2x_1 + 3x_2 - x_3,$$
$$y_2 = 4x_1 + 2x_2 + 7x_3.$$

The object vector

$$\begin{pmatrix} 3 \\ 1 \\ 3 \end{pmatrix}$$

is transformed into the image

$$\begin{pmatrix} 6 \\ 35 \end{pmatrix}$$

by this transformation $Y = AX$ or, equally well, by any other transformation $Z = AU$ with the same matrix A.

We now consider the problem of combining the linear transformation (7.7) represented by the matrix A with a second linear transformation

(7.10) $X = BU,$

where $B = (b_{ij})$ is an $n \times p$ matrix. Clearly, since the y's are expressed in terms of the x's and the x's in terms of the u's, there exists a transformation

(7.11) $Y = CU,$

where C is an $m \times p$ matrix, giving the y's in terms of the u's. To find the exact analytic form of (7.11), we combine (7.7) with the expanded form

$$x_j = \sum_{k=1}^{p} b_{jk} u_k \quad (j = 1, 2, \dots, n)$$

of (7.10), and obtain (for $i = 1, 2, \dots, m$)

$$y_i = \sum_{j=1}^{n} a_{ij} x_j$$

$$= \sum_{j=1}^{n} a_{ij} \sum_{k=1}^{p} b_{jk} u_k$$

$$= \sum_{k=1}^{p} \sum_{j=1}^{n} a_{ij} b_{jk} u_k.$$

Since the expanded form of (7.11) is

$$y_i = \sum_{k=1}^{p} c_{ik} u_k,$$

we see that the coefficients c_{ik} are given by

(7.12) $c_{ik} = \sum_{j=1}^{n} a_{ij} b_{jk} \quad (i = 1, 2, \dots, m; k = 1, 2, \dots, p).$

We state this result as

THEOREM 7.2. If

$$Y = A_{m \times n} X, \quad X = B_{n \times p} U, \quad \text{and} \quad Y = C_{m \times p} U,$$

then the element c_{ik} of C is given by the law

$$c_{ik} = \sum_{j=1}^{n} a_{ij} b_{jk}.$$

The matrix C, which represents the transformation effected by combining the transformations represented by A and B, is called the *product* of A and B; we can thus write

$$Y = AX = A(BU) = (AB)U,$$

or, considering only the matrices involved, $C = AB$. The result (7.12) can be given very neatly as follows:

The element in *row i* and *column k* of the product matrix $C = AB$ is found by applying *row i* of A to *column k* of B and summing all the pairs of products thus obtained.

Note that, with this definition of a matric product, a column vector can *formally* be considered as an $m \times 1$ matrix, and the symbol AX in the linear transformation $Y = AX$ is a true matric product.

EXAMPLE 7.8.

$$A = \begin{pmatrix} 2 & -1 & 3 \\ 4 & 6 & -2 \end{pmatrix}, \qquad B = \begin{pmatrix} 3 & -4 \\ 2 & 0 \\ -1 & 5 \end{pmatrix}.$$

Then

$$AB = \begin{pmatrix} 1 & 7 \\ 26 & -26 \end{pmatrix}.$$

We get the *first* row of AB by applying the first row of A successively to the columns of B; we get the *second* row of AB by applying the *second* row of A successively to the columns of B.

In expanded form, this example reads

$$y_1 = 2x_1 - x_2 + 3x_3, \qquad y_2 = 4x_1 + 6x_2 - 2x_3.$$
$$x_1 = 3u_1 - 4u_2, \qquad x_2 = 2u_1, \qquad x_3 = -u_1 + 5u_2.$$

Then

$$y_1 = 2(3u_1 - 4u_2) - 2u_1 + 3(-u_1 + 5u_2) = u_1 + 7u_2,$$
$$y_2 = 4(3u_1 - 4u_2) + 6(2u_1) - 2(-u_1 + 5u_2) = 26u_1 - 26u_2.$$

EXERCISES

1. Multiply the matrices

$$A = \begin{pmatrix} 2 & 3 & 4 \\ -1 & 0 & 0 \\ 2 & 5 & -2 \end{pmatrix} \quad \text{and} \quad B = \begin{pmatrix} 3 & 7 \\ 2 & 0 \\ 1 & 0 \end{pmatrix}$$

   to obtain AB.

2. Write out the transformations $Y = AX$ and $X = BU$ in Exercise 1, and find AB by actual substitution to give $Y = (AB)U$.

3. For the matrix

$$A = \begin{pmatrix} 2 & -1 \\ 4 & 3 \end{pmatrix},$$

   form A^2, A^3, and A^4.

4. Let A and B be as in Exercise 1; let

$$C = \begin{pmatrix} 2 & 4 \\ 5 & 6 \end{pmatrix}.$$

   Show that $(AB)C$ and $A(BC)$ are equal.

5. Write the transformations

$$y_1 = 2x_1 - 3x_2 \qquad x_1 = 2u_1 + 6u_2$$
$$\text{and}$$
$$y_2 = x_1 + 5x_2 \qquad x_2 = 5u_1 - 2u_2$$

   in matric form, and use matric multiplication to express y_1 and y_2 in terms of u_1 and u_2.

4. PROPERTIES OF MATRICES

Let $A_{m\times n}$ and $B_{n\times p}$ be two matrices; we first note that, in order to form the product AB, the number of columns of A and the number of rows of B must be identical. In general, the product BA will not even exist (unless $m = p$). Even if both AB and BA exist, they will usually differ (even in size).

EXAMPLE 7.9.

$$A = \begin{pmatrix} 3 & -2 & 1 \\ 2 & 4 & 5 \end{pmatrix}, \qquad B = \begin{pmatrix} 2 & 1 \\ 0 & 6 \\ 5 & -2 \end{pmatrix}.$$

Then

$$AB = \begin{pmatrix} 11 & -11 \\ 29 & 16 \end{pmatrix} \quad \text{and} \quad BA = \begin{pmatrix} 8 & 0 & 7 \\ 12 & 24 & 30 \\ 11 & -18 & -5 \end{pmatrix}.$$

Of course, we have no reason for hoping that the commutative law $AB = BA$ should hold for matrix multiplication. Interpretation of the matrices in expanded form as linear transformations makes this clear.

AB	BA
$y_1 = 3x_1 - 2x_2 + x_3$ $y_2 = 2x_1 + 4x_2 + 5x_3$	$z_1 = 2v_1 + v_2$ $z_2 = 6v_2$ $z_3 = 5v_1 - 2v_2$
combines with	combines with
$x_1 = 2u_1 + u_2$ $x_2 = 6u_2$ $x_3 = 5u_1 - 2u_2$	$v_1 = 3w_1 - 2w_2 + w_3$ $v_2 = 2w_1 + 4w_2 + 5w_3$
to give	to give
$y_1 = 11u_1 - 11u_2$ $y_2 = 29u_1 + 16u_2.$	$z_1 = 8w_1 + 7w_3$ $z_2 = 12w_1 + 24w_2 + 30w_3$ $z_3 = 11w_1 - 18w_2 - 5w_3.$

Even in the case that A and B are both square, it is seldom that $AB = BA$, although both products will always exist.

On the other hand, we can readily prove

THEOREM 7.3. The associative law holds for matric multiplication, that is

$$A(BC) = (AB)C.$$

(We assume that the dimensions of A, B, and C are such that the products involved exist.)

Proof. Let $Y = AX$, $X = BU$, $U = CZ$ be the corresponding linear transformations. We first express Y in terms of Z by eliminating X to give

$$Y = (AB)U, \qquad U = CZ,$$

and thence

$$Y = [(AB)C]Z.$$

Then we express Y in terms of Z by eliminating U to give

$$Y = AX, \qquad X = (BC)Z,$$

and thence

$$Y = [A(BC)] Z.$$

Since the expression of Y in terms of Z is independent of whether we get rid of X and then U, or vice versa, we have the result that the two coefficient matrices $A(BC)$ and $(AB)C$ must be identical.

We conclude this section by introducing a useful notation: if c is any constant, then the matrix $cA = c(a_{ij})$ shall denote a matrix in which each element of cA is formed from the corresponding element of A by multiplication by c. For example,

$$c \begin{pmatrix} a_{11} & a_{12} \\ a_{21} & a_{22} \end{pmatrix} = \begin{pmatrix} ca_{11} & ca_{12} \\ ca_{21} & ca_{22} \end{pmatrix},$$

and

$$3 \begin{pmatrix} 5 & 0 & -1 \\ 7 & 4 & 11 \\ -5 & 2 & 2 \end{pmatrix} = \begin{pmatrix} 15 & 0 & -3 \\ 21 & 12 & 33 \\ -15 & 6 & 6 \end{pmatrix}.$$

EXERCISES

1. If

$$A = \begin{pmatrix} 3 & 0 & -2 \\ 4 & 1 & 0 \\ 0 & 2 & 6 \end{pmatrix}, \qquad B = \begin{pmatrix} 2 & 1 & 2 \\ 2 & 3 & 7 \\ 1 & 0 & 0 \end{pmatrix},$$

form AB and BA.

2. Let A be as in Exercise 1. Form AI and IA when

$$I = \begin{pmatrix} 1 & 0 & 0 \\ 0 & 1 & 0 \\ 0 & 0 & 1 \end{pmatrix}.$$

3. Let

$$A = \begin{pmatrix} x & y \\ -y & x \end{pmatrix}, \qquad B = \begin{pmatrix} u & v \\ -v & u \end{pmatrix};$$

form AB and BA.

4. Prove Theorem 7.3 by forming the element in row i and column k of $(AB)C$ and of $A(BC)$ from the analytic expressions,

$$y_i = \Sigma a_{ij} x_j, \quad x_j = \Sigma b_{jm} u_m, \quad u_m = \Sigma c_{mk} z_k.$$

In Exercises 5 and 6, the matrices A and B are $n \times n$ matrices.

5. Prove that the sum of the diagonal elements of BA is the same as the sum of the diagonal elements of AB.

6. Let $B = cA$; prove that det $B = c^n$ det A, where det A denotes the $n \times n$ determinant formed from the n^2 entries in A.

5. SQUARE MATRICES

From this point onward, we confine ourselves to the only case of great practical importance, that in which the matrices are square. Our transformation $Y = AX$ then expresses n quantities $y_1, \ldots, y_n$ in terms of n quantities $x_1, \ldots, x_n$, and it becomes a natural question to ask whether the x's can be expressed in terms of the y's. The answer is given by

THEOREM 7.4. *If $Y = AX$, then we can find a unique inverse transformation $X = BY$ if the determinant D formed from the entries in A is non-zero.*

Proof. Let $Y = AX$ be written as

$$y_i = \sum_{j=1}^{n} a_{ij} x_j \quad (i = 1, \ldots, n),$$

and let $D = \det A = |a_{ij}|$; let D_j be D with its jth column replaced by $y_1, y_2, \ldots, y_n$. Then, by Cramer's Theorem, x_j is given uniquely by

$$x_j = \frac{D_j}{D}.$$

Expanding along the jth column of D_j gives

$$x_j = \frac{1}{D} \sum_{i=1}^{n} y_i A_{ij},$$

$$x_j = \sum_{i=1}^{n} \frac{A_{ij}}{D} y_i \quad (j = 1, 2, \ldots, n).$$

This result gives $X = BY$, where the matrix B is uniquely defined by

$$(7.13) \qquad B = \frac{1}{D} \begin{vmatrix} A_{11} & A_{21} & A_{31} & \cdots & A_{n1} \\ A_{12} & A_{22} & A_{32} & \cdots & A_{n2} \\ \cdots & \cdots & \cdots & \cdots & \cdots \\ A_{1n} & A_{2n} & A_{3n} & \cdots & A_{nn} \end{vmatrix}.$$

Certainly B can be found if $D = \det A \neq 0$. This matrix B which expresses the x's in terms of the y's is called "A inverse" and is denoted by A^{-1}. We see that the element in *row i* and *column j* is found by taking the cofactor A_{ji} and dividing by D.

EXAMPLE 7.10.

$$A = \begin{pmatrix} 3 & -2 & 6 \\ 4 & 1 & 1 \\ 2 & 1 & -2 \end{pmatrix}.$$

We have

$$\begin{aligned} A_{11} &= -3, & A_{12} &= 10, & A_{13} &= 2, \\ A_{21} &= 2, & A_{22} &= -18, & A_{23} &= -7, \\ A_{31} &= -8, & A_{32} &= 21, & A_{33} &= 11. \end{aligned}$$

$$\det A = 3 \begin{vmatrix} 1 & 1 \\ 1 & -2 \end{vmatrix} + 2 \begin{vmatrix} 4 & 1 \\ 2 & -2 \end{vmatrix} + 6 \begin{vmatrix} 4 & 1 \\ 2 & 1 \end{vmatrix} = -17.$$

Consequently,

$$A^{-1} = \begin{pmatrix} \frac{3}{17} & -\frac{2}{17} & \frac{8}{17} \\ -\frac{10}{17} & \frac{18}{17} & -\frac{21}{17} \\ -\frac{2}{17} & \frac{7}{17} & -\frac{11}{17} \end{pmatrix}.$$

Closely connected with the idea of the *inverse* transformation is that of the *identity* transformation $y_i = x_i$ ($i = 1, 2, \dots, n$). Its matrix

$$\begin{pmatrix} 1 & 0 & 0 & \dots & 0 \\ 0 & 1 & 0 & \dots & 0 \\ 0 & 0 & 1 & \dots & 0 \\ & & \cdots & \cdots & \\ 0 & 0 & 0 & \dots & 1 \end{pmatrix}$$

is denoted by the symbol I_n (usually written simply as I when the dimension of the matrix is obvious). By compounding the transformations $Y = IX = X$ and $X = AU$, we obtain

$$Y = IAU = AU.$$

Similarly, the transformations $Y = AX$ and $X = IU = U$ give

$$Y = AIU = AU.$$

This result is expressed matrically by equation

(7.14) $AI = IA = A,$

that is, *the identity matrix multiplies every other matrix to leave it unaltered.* Further, if we consider the inverse transformations $Y = AX$ and $X = A^{-1}Y$, we at once obtain

$$Y = A(A^{-1}Y) = AA^{-1}Y.$$

But $Y = IY$, and hence $AA^{-1} = I$; similarly, $X = A^{-1}(AX) = X$, and we obtain

(7.15) $AA^{-1} = A^{-1}A = I.$

We conclude this section by putting Cramer's Rule in matric form; let

$$AX = C$$

represent n linear equations in n unknowns, as in (7.6). Multiply by A^{-1} on the left, and we obtain

$$A^{-1}AX = A^{-1}C,$$
$$IX = A^{-1}C,$$
(7.16) $$X = A^{-1}C.$$

Formula (7.16) is Cramer's Rule. Thus we can always solve the linear system $AX = C$ if we know the inverse matrix A^{-1}.

As an illustration of Equation (7.16), let us consider

$$y_1 = 2x_1 + 3x_2, \quad y_2 = 3x_1 - 5x_2;$$

we eliminate x_2 and obtain

$$5y_1 + 3y_2 = 19x_1,$$

whence

$$x_1 = \tfrac{5}{19}y_1 + \tfrac{3}{19}y_2;$$

by substitution,

$$x_2 = \tfrac{3}{19}y_1 - \tfrac{2}{19}y_2.$$

This elimination procedure has solved the equations for X in terms of Y; it has also produced the inverse matrix

$$\begin{pmatrix} 2 & 3 \\ 3 & -5 \end{pmatrix}^{-1} = \begin{pmatrix} \tfrac{5}{19} & \tfrac{3}{19} \\ \tfrac{3}{19} & -\tfrac{2}{19} \end{pmatrix}.$$

Thus, if we are given a system of equations, we can alternatively state that we are really given a matrix of coefficients; conversely, if we are given a matrix, it is equivalent to our being given a system of equations.

EXERCISES

1. (a) Given

$$A = \begin{pmatrix} 7 & 3 \\ 11 & 5 \end{pmatrix},$$

compute A^{-1} and check that $AA^{-1} = A^{-1}A = I$.

(b) Repeat Exercise 1(a) if

$$A = \begin{pmatrix} 2 & 0 & -3 \\ 6 & 4 & 5 \\ 7 & 1 & 0 \end{pmatrix}.$$

2. Let A be as in Exercise 1(b); use A^{-1} to give the solution of the equations

$$2x_1 \quad - \quad 3x_3 = \quad 4,$$
$$6x_1 + 4x_2 + 5x_3 = \quad 1,$$
$$7x_1 + \ x_2 \qquad = -5.$$

3. Write the equations

$$5x_1 - 2x_2 + \ x_3 = 4,$$
$$7x_1 + \ x_2 - 5x_3 = 8,$$
$$3x_1 + 7x_2 + 4x_3 = 10,$$

in matric form. Compute A^{-1}, and thence deduce the solution of the equations.

4. Let

$$A = \begin{pmatrix} 7 & 1 & 2 \\ 2 & 2 & 4 \\ -1 & 5 & 1 \end{pmatrix}, \quad C = \begin{pmatrix} 2 \\ 1 \\ 3 \end{pmatrix}.$$

Form the product $A^{-1}C$; what equations have you thereby solved?

5. It can be established that det AB = det A det B, that is, that the determinant of a product is the product of the determinants. Assuming this result, show that (Note: A is called *singular* if det A = 0.)

(a) if $AB = I$, then neither A nor B can be singular;
(b) if $Y = AX$, with A singular, then it is not possible to find a matrix B such that $X = BY$.

6. Prove that $(AB)^{-1} = B^{-1}A^{-1}$.

7. Find the inverse of

$$A = \begin{pmatrix} 5.0 & 7.6 & -3.2 \\ 3.1 & -8.5 & 6.4 \\ -9.2 & 5.3 & 2.7 \end{pmatrix},$$

giving the entries of A^{-1} to five decimals.

8. Let A^* be the *approximate* inverse (rounded to five decimals) obtained in Exercise 7. Compute AA^* and A^*A, verifying that they are not equal (because of round-off errors, $A^* \neq A^{-1}$, and $AA^* \neq AA^{-1}$).

9. Use the value A^* obtained in Exercise 7 to give an approximate solution of $AX = C$, where C is the column vector with components 18.2, 31.8, −5.2.

10. Find all matrices of the form

$$\begin{pmatrix} 0 & x \\ y & z \end{pmatrix}$$

which commute with

$$\begin{pmatrix} 3 & 7 \\ -2 & 0 \end{pmatrix}.$$

11. Given

$$A = \begin{pmatrix} 3 & 1 & 2 \\ -1 & 0 & 5 \\ 1 & 1 & 2 \end{pmatrix}, \qquad C = \begin{pmatrix} 5 \\ 9 \\ -7 \end{pmatrix},$$

solve the equations $AX = C$ by finding A^{-1}.

12. Prove that the sum of the diagonal elements of A is the same as the sum of the diagonal elements of $B^{-1}AB$.

13. If $AB = I$, prove that $BA = I$.

6. THE DIRECT METHOD OF COMPUTING THE INVERSE MATRIX

If a matrix A is larger than 3×3, the computation of A^{-1} by cofactors becomes onerous, and other methods are generally advisable. We shall describe the simplest of these.

Let us define elementary row operations ϕ_i on a matrix A as being any of the following operations:

(a) multiplication of any row by a non-zero constant;

(b) addition to any row of a multiple of any other row;

(c) interchange of any two rows.

Consider the linear transformation $Y = AX$ written in the form $AX = IY$. Suppose that $\phi_1, \phi_2, \ldots, \phi_n$ is a sequence of elementary row operations which transforms A into the identity I. Then

$$\phi_n \ldots \phi_2\phi_1 AX = \phi_n \ldots \phi_2\phi_1 IY;$$

by assumption, $\phi_n \ldots \phi_2\phi_1 A = I$. Hence

$$X = \phi_n \ldots \phi_2\phi_1 IY.$$

But $X = A^{-1}Y$; hence

$$A^{-1} = \phi_n \ldots \phi_2\phi_1 I.$$

Thus we have

THEOREM 7.5. If a sequence of elementary row operations will transform A into I, then this *same* sequence of elementary row operations will transform I into A^{-1}.

Actually, as we shall see in the exercises, the ϕ_i can themselves be represented as matrices. We shall show how the ϕ_i can always be systematically constructed by working out

EXAMPLE 7.11.

$$A = \begin{pmatrix} 5.0 & 7.6 & -3.2 \\ 3.1 & -8.5 & 6.4 \\ -9.2 & 5.3 & 2.7 \end{pmatrix}.$$

The method is almost self-explanatory (Table 7.1); the first three columns of Table 7.1 give the elements of A and the results of applying elementary row operations ϕ_i; the second three columns give the elements of I and the results subsequent to applying the *same* operations ϕ_i; the seventh column is a check column made up of the sums of all elements in the various rows and the results of applying ϕ_i to these sums.

For greater ease in explanation, we stretch the work into five steps, separated by horizontal lines. In step 1 we divide the first row of A by 5.0; this makes $a_{11} = 1$, and we subtract multiples of this first row from the subsequent rows in order to get zeros in the remainder of the first column. In step 2 we make $a_{22} = 1$, and in step 3 we get zeros in all remaining positions in the second column. In step 4 we make $a_{33} = 1$; finally, in step 5 we get zeros in all the rest of the third column. The table follows.

Table 7.1. DIRECT METHOD OF COMPUTING THE INVERSE MATRIX

5.0	7.6	−3.2	1	0	0	10.4
3.1	−8.5	6.4	0	1	0	2.0
−9.2	5.3	2.7	0	0	1	−0.2
1	1.52	−.64	.20	0	0	2.08
0	−13.212	8.384	−.62	1	0	−4.448
0	19.284	−3.188	1.84	0	1	18.936
1	1.52	−.64	.20	0	0	2.08
0	1	−.63457	.04693	−.07569	0	.33667
0	19.284	−3.188	1.84	0	1	18.936
1	0	.32455	.12867	.11505	0	1.56826
0	1	−.63457	.04693	−.07569	0	.33667
0	0	9.04905	.93500	1.45961	1	12.44366
1	0	.32455	.12867	.11505	0	1.56826
0	1	−.63457	.04693	−.07569	0	.33667
0	0	1	.10333	.16130	.11051	1.37513
1	0	0	.09513	.06270	−.03587	1.12196
0	1	0	.11250	.02667	.07013	1.20929
0	0	1	.10333	.16130	.11051	1.37513

The resulting approximation to A^{-1} (an approximation, since only five decimals were retained in the work) is given by

$$A^{-1} \doteq \begin{pmatrix} .09513 & .06270 & -.03587 \\ .11250 & .02667 & .07013 \\ .10333 & .16130 & .11051 \end{pmatrix}.$$

With practice the number of steps in the process can be greatly reduced. The method has the advantage of great straightforwardness; it can hardly be forgotten or abused. There are other more elegant methods for finding A^{-1}, but their complexity renders them unsuitable for casual or occasional use.

EXERCISES

1. Apply the method of this section to compute A^{-1}, given

$$A = \begin{pmatrix} 2.5 & -3.8 \\ 6.7 & 5.4 \end{pmatrix}. \quad \text{Keep a check column.}$$

2. Let $R_i = R_i(\lambda)$ be the matrix obtained from I by multiplying the ith row of I by λ. Show that R_iA is a matrix which is obtained from A by multiplying the ith row of A by λ (in other words, R_i performs the first type of elementary row operation on A).

3. Let $F_{ij} = F_{ij}(\lambda)$ be the matrix obtained from I by multiplying the jth row of I by λ and adding it to the ith row. Show that $F_{ij}A$ is a matrix which is obtained from A by adding λ times the jth row of A to the ith row (that is, F_{ij} performs the second type of elementary row operation on A).

4. Let P_{ij} be the matrix obtained from I by interchanging the ith and jth rows of I. Show that $P_{ij}A$ is a matrix which is obtained from A by interchanging the ith and jth rows of A (that is, P_{ij} performs the third type of elementary row operation on A).

5. Supposing A to be non-singular, describe a systematic method for reducing A to the identity I by premultiplication by matrices of the form R_i and F_{ij} (follow the method of Example 7.11). Show that, in general, n operations of the type R_i and at most n^2 of the type F_{ij} are required by this procedure.

6. Prove:
$$[R_i(\lambda)]^{-1} = R_i(\lambda^{-1}),$$
$$[F_{ij}(\lambda)]^{-1} = F_{ij}(-\lambda),$$
$$[P_{ij}]^{-1} = P_{ij}.$$

7. If A is non-singular, deduce from Exercise 5 that A can be written as a product of matrices of the types R_i and F_{ij}.

8. Prove $\quad\quad \det(R_iA) = \det R_i \det A,$
and $\quad\quad \det(F_{ij}A) = \det F_{ij} \det A.$

9. Deduce from Exercises 7 and 8 that if A is non-singular then $\det(AB) = \det A \det B$. (This result also holds for A singular.)

10.
$$A = \begin{pmatrix} 3 & 1 \\ -2 & 5 \end{pmatrix}, \quad B = \begin{pmatrix} 2 & 9 \\ 8 & -4 \end{pmatrix}.$$

Compute AB, and verify the result of Exercise 9 by finding $\det A$, $\det B$, and $\det AB$.

11. Find the inverse of

$$A = \begin{pmatrix} 2.4 & 6.7 & 3.2 \\ -5.8 & 7.2 & 8.1 \\ -1.9 & -5.6 & 6.4 \end{pmatrix}$$

by the method of Example 7.11 (carry four decimals throughout).

12. Use the method of Example 7.11 to find (to five decimals) the inverse of

$$\begin{pmatrix} 2.4 & 5.2 \\ -6.5 & 7.6 \end{pmatrix}.$$

Check by forming AA^{-1} and $A^{-1}A$.

13. Obtain A^{-1} in Exercise 12 by performing the same operations used there upon the equations

$$y_1 = 2.4x_1 + 5.2x_2,$$
$$y_2 = -6.5x_1 + 7.6x_2,$$

and thus directly expressing x_1 and x_2 in terms of y_1 and y_2.

14. Let A' denote the transposed matrix in which the rows of A become the columns of A' (the element $a'_{ij} = a_{ji}$). Prove that $(AB)' = B'A'$.

15. Prove det $A = $ det $B^{-1}AB$.

16. Let

$$A = \begin{pmatrix} 1 & 2 \\ 1000 & 2001 \end{pmatrix}, \qquad B = \begin{pmatrix} 2000.99 & -2 \\ -1000 & 1 \end{pmatrix}.$$

Compute AB and BA. Note that B is a poor right inverse for A; what is the exact value of A^{-1}?

17. Let

$$A = \begin{pmatrix} 100 & 101 \\ 99 & 100 \end{pmatrix}, \qquad B = \begin{pmatrix} 99.99 & -98.99 \\ -98.99 & 98.01 \end{pmatrix}.$$

Compare $AB - I$ and $BA - I$.

18. Invert the matrix

$$\begin{pmatrix} 5.07 & 3.19 & 2.32 & 1.87 \\ 1.21 & 8.95 & -2.04 & 4.49 \\ -3.22 & -4.66 & 9.93 & 2.05 \\ 1.18 & 6.72 & -3.66 & 8.77 \end{pmatrix}.$$

7. EIGENVALUE PROBLEMS

In connection with matrices, there are two main numerical problems which arise. We have already discussed one of these, namely, the problem of finding A^{-1}, given A. The other problem, which occurs very frequently in applied mathematics, is: "Given A, find a vector X and a constant λ such that $AX = \lambda X$, that is, such that the matrix A transforms X into a multiple of itself".

If X and λ have been so found, we say that X is an eigenvector of the transformation A; λ is called an eigenvalue of A. We at once prove

LEMMA 7.7. If X is an eigenvector with corresponding eigenvalue λ, then aX is an eigenvector for any $a \neq 0$.

Proof.

$$A(aX) = aAX = a\lambda X = \lambda(aX).$$

Thus, an eigenvector X is determined only up to a constant of proportionality.

We also have

LEMMA 7.8. There are n eigenvalues (not necessarily all distinct) for an $n \times n$ matrix A.

Proof. Let us consider the n equations represented by $AX = \lambda X$. We write $AX - \lambda IX = 0$, or equivalently $(A - \lambda I)X = 0$.

In this form, we have n homogeneous linear equations in n unknowns; using the Kronecker δ, these equations may be expressed as

$$\sum_{j=1}^{n} (a_{ij} - \lambda \delta_{ij})x_j = 0 \quad (i = 1, 2, \ldots, n);$$

in order for these equations to have non-trivial solutions for the x's, we must have

$$\det (A - \lambda I) = \det (a_{ij} - \lambda \delta_{ij}) = 0.$$

This is an nth degree equation to determine λ, and consequently there will be n eigenvalues $\lambda_1, \lambda_2, \ldots, \lambda_n$.

Before working on an example, it might be well to mention some physical situations where eigenvalues are required, since eigenvalues always seem somewhat artificial to a student when he first encounters them. Such situations include finding the fundamental frequencies associated with:

    (a) the oscillation of a triple pendulum,
    (b) the torsional oscillations of a uniform cantilever,
    (c) the torsional oscillations of a multi-cylinder engine,
    (d) the flexural oscillations of a tapered beam,
    (e) the symmetric vibrations of an annular membrane,
    (f) the oscillations of an aeroplane wing in an airstream,
    (g) the vibration of a system of masses and springs,
    (h) the buckling of a structure of spring-supported rigid links.

For complete formulation of the preceding physical situations, the student is referred to Frazer, Duncan, and Collar, *Elementary Matrices* (pp. 310–331) and Crandall, *Engineering Analysis* (pp. 61–67). Once the physical formulation is completed, however, all these problems have the same mathematical form, namely, to find an eigenvalue for a numerical matrix as in

EXAMPLE 7.12. Find the eigenvalues of the matrix

$$A = \begin{pmatrix} 3 & 2 & 5 \\ 6 & -5 & 3 \\ -24 & 38 & 2 \end{pmatrix}.$$

The equation determining the λ_i is

$$\begin{vmatrix} 3-\lambda & 2 & 5 \\ 6 & -5-\lambda & 3 \\ -24 & 38 & 2-\lambda \end{vmatrix} = 0,$$

$$(3-\lambda)\begin{vmatrix} -5-\lambda & 3 \\ 38 & 2-\lambda \end{vmatrix} - 2\begin{vmatrix} 6 & 3 \\ -24 & 2-\lambda \end{vmatrix} + 5\begin{vmatrix} 6 & -5-\lambda \\ -24 & 38 \end{vmatrix} = 0,$$

$$\lambda^3 - 25\lambda = 0.$$

Hence

$$\lambda_1 = 0, \quad \lambda_2 = 5, \quad \lambda_3 = -5.$$

Many of the matrices one meets in practice are *symmetric*, that is, $a_{ij} = a_{ji}$. For these matrices we have the important

THEOREM 7.6. All eigenvalues of a symmetric matrix with real elements are real.

Proof. Let λ be an eigenvalue of the real symmetric matrix A. Let X be a corresponding eigenvector; let A' denote the transposed matrix (that is, the rows of A are the columns of A').

We have $AX = \lambda X$. Multiply by the $1 \times n$ matrix $\bar{X}'$, where the bar denotes that all elements of $\bar{X}'$ are the complex conjugates of those of X'. We then have

$$\bar{X}'AX = \lambda\bar{X}'X.$$

Taking the conjugate complex of this number yields

$$X'\bar{A}\bar{X} = \bar{\lambda}X'\bar{X},$$

or simply

$$X'A\bar{X} = \bar{\lambda}X'\bar{X},$$

since $\bar{A} = A$ (A being real). Now using the result $(AB)' = B'A'$ (cf. Exercise 14 of Section 6), we take transposes of both sides of the equation to obtain

$$\bar{X}'A'X = \bar{\lambda}\bar{X}'X.$$

Since A is symmetric, $A' = A$, and we have

$$\bar{X}'AX = \bar{\lambda}\bar{X}'X = \lambda\bar{X}'X.$$

Now

$$\bar{X}'X = \sum_{j=1}^{n} \bar{x}_j x_j = \text{a positive real number.}$$

Hence $\bar{\lambda} = \lambda$, that is, λ is real.

COROLLARY 1. X is a vector with real components.

COROLLARY 2. If λ_1 and λ_2 are distinct eigenvalues of a real symmetric matrix A, and if X_1 and X_2 are the corresponding eigenvectors with components x_{1j} and x_{2j} respectively, then

$$\sum_{j=1}^{n} x_{1j} x_{2j} = X_1' X_2 = 0.$$

(We refer to this property by saying that X_1 and X_2 are *orthogonal* vectors.)

Proof. $AX_1 = \lambda_1 X_1$ implies $X_1' A' = X_1' A = \lambda_1 X_1'$.

Hence $\qquad\qquad\qquad X_1' A X_2 = \lambda_1 X_1' X_2.$

Also $\qquad\qquad AX_2 = \lambda_2 X_2$ implies $X_1' A X_2 = \lambda_2 X_1' X_2.$

Hence we have $\qquad\quad X_1' A X_2 = \lambda_1 X_1' X_2 = \lambda_2 X_1' X_2.$

We then have $\qquad\qquad (\lambda_1 - \lambda_2) X_1' X_2 = 0.$

But $\lambda_1 - \lambda_2 \neq 0$ by hypothesis. Hence

$$X_1' X_2 = \sum_{j=1}^{n} x_{1j} x_{2j} = 0.$$

EXERCISES

1. Let A be the matrix

$$\begin{pmatrix} 3 & 2 & -1 \\ 5 & 0 & 6 \\ 2 & 3 & 1 \end{pmatrix}.$$

(a) Compute A^{-1}.

(b) Let B be the vector

$$\begin{pmatrix} 2 \\ -5 \\ 8 \end{pmatrix};$$

form the product $A^{-1}B$. What set of equations have you thereby solved?

(c) Find the eigenvalues of A correct to three decimals.

(d) Obtain the corresponding eigenvectors.

2. Let $\lambda_1, \lambda_2, \ldots, \lambda_n$ be the n eigenvalues of A. Prove:

(a) $$\sum_{i=1}^{n} \lambda_i = \sum_{i=1}^{n} a_{ii},$$

(b) $$\prod_{i=1}^{n} \lambda_i = \det A.$$

3. Obtain the eigenvalues of the following matrix.

$$\begin{pmatrix} 1 & 1 & -2 & 0 \\ 1 & 2 & 0 & 5 \\ -2 & 0 & 3 & 4 \\ 0 & 5 & 4 & 10 \end{pmatrix}$$

Find the eigenvector corresponding to the largest eigenvalue.

4. Let

$$A = \begin{pmatrix} 1 & 3 & -1 \\ 3 & 2 & 4 \\ -1 & 4 & 10 \end{pmatrix}.$$

Find all three eigenvalues.

5. Find the eigenvalues of the matrix with elements a_{ii} on the diagonal and zeros everywhere else.

6. If $\lambda_1 \neq \lambda_2$, then the corresponding eigenvectors X_1 and X_2 differ (assume they are the same, and derive a contradiction).

7. Let R' be a row vector with the property that $R'A = \mu R'$; prove that there are n possible values for the constant μ, and that these are just the n eigenvalues of A.

8. ITERATIVE DETERMINATION OF EIGENVALUES

Let A be a given matrix whose eigenvalues are required. Let X_1 be an arbitrary vector, which we use as an approximation to an eigenvector; suppose we "normalize" X_1 by requiring that one component, say the last, be unity. Compute the sequence

$$AX_1 = \lambda_1 X_2, \quad AX_2 = \lambda_2 X_3, \quad AX_3 = \lambda_3 X_4, \quad \ldots .$$

In this sequence, all the vectors X_i are to be normalized in whatever manner was chosen originally. Then, if the X_i approach a limit X, the recursion relation will approach the limiting relation

$$AX = \lambda_i X,$$

and we see that the constant λ_i will also approach a limit λ. The relation approached by the successive iterates will thus be

$$AX = \lambda X;$$

thus $\lambda = \lim \lambda_i$ is an eigenvalue of A, and $X = \lim X_i$ is the corresponding eigenvector of A.

In general, this process will converge to give the largest eigenvalue of A provided that this eigenvalue is real and unrepeated; in most physical problems, this may be the only eigenvalue required.

EXAMPLE 7.13.

$$A = \begin{pmatrix} 1 & 3 & -1 \\ 3 & 2 & 4 \\ -1 & 4 & 10 \end{pmatrix}.$$

Since we have no information, let us select

$$X_1 = \begin{pmatrix} 0 \\ 0 \\ 1 \end{pmatrix}.$$

Then

$$AX_1 = \begin{pmatrix} -1 \\ 4 \\ 10 \end{pmatrix} = 10\begin{pmatrix} -.1 \\ .4 \\ 1 \end{pmatrix} = 10X_2,$$

$$AX_2 = \begin{pmatrix} .1 \\ 4.5 \\ 11.7 \end{pmatrix} = 11.7\begin{pmatrix} .009 \\ .385 \\ 1 \end{pmatrix} = 11.7X_3,$$

$$AX_3 = \begin{pmatrix} .164 \\ 4.797 \\ 11.531 \end{pmatrix} = 11.531\begin{pmatrix} .014 \\ .416 \\ 1 \end{pmatrix} = 11.531X_4,$$

$$AX_4 = \begin{pmatrix} .262 \\ 4.874 \\ 11.650 \end{pmatrix} = 11.650\begin{pmatrix} .022 \\ .418 \\ 1 \end{pmatrix} = 11.650X_5,$$

$$AX_5 = \begin{pmatrix} .276 \\ 4.902 \\ 11.650 \end{pmatrix} = 11.650\begin{pmatrix} .024 \\ .421 \\ 1 \end{pmatrix} = 11.650X_6,$$

$$AX_6 = \begin{pmatrix} .287 \\ 4.914 \\ 11.660 \end{pmatrix} = 11.660\begin{pmatrix} .025 \\ .421 \\ 1 \end{pmatrix} = 11.660X_7,$$

$$AX_7 = \begin{pmatrix} .288 \\ 4.917 \\ 11.659 \end{pmatrix} = 11.659\begin{pmatrix} .025 \\ .422 \\ 1 \end{pmatrix} = 11.659X_8,$$

$$AX_8 = \begin{pmatrix} .291 \\ 4.919 \\ 11.663 \end{pmatrix} = 11.663\begin{pmatrix} .025 \\ .422 \\ 1 \end{pmatrix} = 11.663X_9.$$

The limit has now been reached; we find an eigenvalue of 11.66 (the third decimal being uncertain) and a corresponding eigenvector with components .025, .422, and 1.

EXERCISES

1. Use iteration to find an eigenvalue and eigenvector for the matrix

$$A = \begin{pmatrix} 2 & -1 & 2 \\ -1 & 6 & 8 \\ 2 & 8 & 30 \end{pmatrix}.$$

Check by finding the equation $\det (A - \lambda I) = 0$ and solving it by Newton's method.

2. Why is

$$\begin{pmatrix} 0 \\ 1 \\ 0 \\ 0 \end{pmatrix}$$

a good initial guess from which to start the iterative procedure for

$$\begin{pmatrix} 2 & 3 & 0 & 4 \\ 3 & 20 & 3 & -5 \\ 0 & 3 & 1 & 0 \\ 4 & -5 & 0 & 1 \end{pmatrix}?$$

Carry out the iteration.

3. Starting from

$$\begin{pmatrix} 1 \\ 0 \\ 0 \end{pmatrix},$$

find an eigenvalue of

$$\begin{pmatrix} 25 & 1 & 2 \\ 1 & 3 & 0 \\ 2 & 0 & -4 \end{pmatrix}.$$

4. Repeat Exercise 3 with

$$\begin{pmatrix} 1 \\ 1 \\ 1 \end{pmatrix}$$

as initial vector.

5. Show that

$$\begin{pmatrix} 2.5 & 3.1 & 1.6 \\ 3.1 & 8.9 & 7.2 \\ 1.6 & 7.2 & 23.8 \end{pmatrix}$$

has dominant eigenvalue 26.99913 with corresponding eigenvector

$$\begin{pmatrix} .11821 \\ .41806 \\ 1 \end{pmatrix}.$$

6. Find the largest eigenvalue of

$$\begin{pmatrix} 5 & 2 & 1 & -2 \\ 2 & 6 & 3 & -4 \\ 1 & 3 & 19 & 2 \\ -2 & -4 & 2 & 1 \end{pmatrix};$$

find also the corresponding eigenvector.

7. Carrying six decimals, extend Example 7.13 as far as the stage of finding that the eigenvector has components .024867, .421714, 1, and that the corresponding eigenvalue is 11.66199 (rounded from 11.661989).

8. Find the largest eigenvalue and the associated eigenvector of

$$\begin{pmatrix} 6 & 2 & 0 & -1 & 5 \\ 2 & 13 & 4 & 0 & -2 \\ 0 & 4 & 0 & -1 & -1 \\ -1 & 0 & -1 & 5 & 0 \\ 5 & -2 & -1 & 0 & -2 \end{pmatrix}.$$

9. Use an iterative procedure to find the largest eigenvalue and the corresponding eigenvector of the matrix

$$\begin{pmatrix} 3 & 2 & 6 \\ -1 & 12 & 1 \\ 4 & 2 & 1 \end{pmatrix}$$

(carry four decimals throughout).

10. Prove that the eigenvalue with least absolute value of A can be found as the eigenvalue with greatest absolute value of A^{-1}.

CHAPTER EIGHT

Solution of Linear Equations

1. THE METHOD OF EXACT ELIMINATION

If the number of equations is not too great, and the coefficients are not too unwieldy, the method of exact elimination possesses many advantages. It produces the answer in finite form; it avoids the introduction of round-off errors; and, finally, it avoids the difficulties other methods encounter if the system of equations is "ill-conditioned" (see Section 6). Two examples should make the method clear.

EXAMPLE 8.1.

$$6x + 8y - 7z = 8 \qquad (1),$$
$$10x + 5y + 8z = 4 \qquad (2),$$
$$7x - 9y + 7z = 11 \qquad (3).$$

Combining first equations (1) and (2), and then (1) and (3), we obtain

$$25y - 59z = 28 \qquad (4),$$
$$125y - 14z = -82 \qquad (5).$$

Now eliminate y to give

$$-21z = 222 \qquad (6).$$

Solving (6), and substituting in (4) and (1), gives

$$z = -\frac{222}{281}, \quad y = -\frac{1046}{1405}, \quad x = \frac{1973}{1405}.$$

Decimally, $x = 1.4043$, $y = -.7445$, $z = -.7900$.

EXAMPLE 8.2.

$$3.7x + 5.2y + 9.1z - 2.8u = \quad 5.1 \qquad (1),$$
$$2.5x - 3.8y + 7.4z + 8.7u = \quad 8.2 \qquad (2),$$
$$9.5x + 4.4y + 6.8z + 3.0u = \quad 13.4 \qquad (3),$$
$$1.6x - 1.9y - 2.4z - 7.5u = -9.8 \qquad (4).$$

Combine (1) successively with (2), (3), and (4), after multiplying each equation by 10 to remove decimals.

$$2706y - \quad 463z - 3919u = -1759 \qquad (5),$$
$$3312y + 6129z - 3770u = \quad -113 \qquad (6),$$
$$1535y + 2344z + 2327u = \quad 4442 \qquad (7).$$

Now combine (5) successively with (6) and (7).

$$18118530z + 2778108u = 5520030 \qquad (8),$$
$$7053569z + 12312527u = 14720117 \qquad (9).$$

Finally, eliminating z,

$$203489313357858u = 227770968980940 \qquad (10).$$

Solving, from (10), (8), (5), and (1) gives

$$u = 1.11932644, \quad z = .13303619,$$
$$y = \quad .99380490, \quad x = .50154033.$$

As a check, substitute in (2), (3), and (4). The left-hand sides are respectively

$$8.200000039, \quad 13.400000107, \quad \text{and} \quad -9.799999938.$$

This verifies the solution.

We have stressed this method of elimination because, since it is taught at the secondary-school level, the misconception commonly prevails that it is plebeian and should be abandoned in favour of determinants or some other "elegant" method as soon as the student is capable of such a refinement. Exactly the opposite is the case; the method of elimination is capable of yeoman service in a great many situations; we have already discussed the virtues of the method of determinants in the previous chapter.

EXERCISES

1. Solve the equations

$$4.96x - 5.37y = 6.92,$$
$$3.92x + 4.85y = 2.14.$$

2. Solve the equations

$$3x - 2y + 5z = 2,$$
$$4x + y + 2z = 4,$$
$$2x - y + 4z = 7.$$

3. Solve the equations

$$2.3x + 4.1y - 2.7z = 8.5,$$
$$1.4x + 1.9y + 4.5z = 7.8,$$
$$2.9x - 4.5y + 3.6z = 10.3.$$

4. Solve the equations

$$4x - 2y + 5z + u = 8,$$
$$2x + 3y + 6z - 2u = 7,$$
$$3x + 5y + 9z + 5u = 10,$$
$$2x - 7y + 3z + 4u = 7.$$

2. THE METHOD OF TRIANGULAR ELIMINATION

Example 8.2 pointed up clearly that if the coefficients had been much larger initially, or if the number of equations involved had been greater, the numerical labour would have increased to unmanageable proportions. When such is the case, we abandon the attempt to obtain equations exactly equivalent to the original set [as (1), (5), (8), and (10) in Example 8.2], and work to a certain number of decimals only. This procedure has the effect of replacing the original set of equations by an approximately equivalent set. If the equations are well-conditioned, that is, if the solutions are not unduly sensitive to minute changes in the coefficients, we end up with an approximate solution. In order to minimize round-off errors, we retain, at any stage, the maximum coefficient in the whole set of equations. The process is exemplified in

EXAMPLE 8.3.

$$4.405x + 2.972y - 3.954z = 8.671 \qquad (1),$$
$$3.502x + \underline{9.617y} - 2.111z = 12.345 \qquad (2),$$
$$7.199x + 3.248y + 1.924z = 6.227 \qquad (3).$$

Rewrite (2) and eliminate y from (1) and (3). Retain three decimals in the work.

$$9.617y + 3.502x - 2.111z = 12.345 \qquad (4),$$
$$31.955x - 31.752z = 46.700 \qquad (5),$$
$$\underline{57.858x} + 25.360z = 19.788 \qquad (6).$$

Retain (6) and eliminate x to give the final form

$$9.617y + 3.502x - 2.111z = \quad 12.345 \qquad (4),$$
$$57.858x + 25.360z = \quad 19.788 \qquad (6),$$
$$-2647.486z = 2069.643 \qquad (7).$$

From this final triangular array, we end up with $z = -.7817$, $x = .6846$, $y = .8628$. Substituting this solution in (1) and (3), we obtain 8.671 and 6.227, thus verifying the solution.

As can be seen from this example, the method is really just a special modification of the method of exact elimination. Under most conditions, the method of this section is more convenient.

EXERCISES

1. Solve
$$3.246x - 9.542y = 7.851,$$
$$5.921x + 6.512y = 22.127.$$

2. Solve
$$3.512x + 5.617y - 9.222z = 11.314,$$
$$8.122x + 6.914y + 2.776z = 6.528,$$
$$1.443x - 2.730y + 13.020z = 15.917.$$

3. Solve
$$4.942x + 3.619y + 2.872z = 3.007,$$
$$2.193x + 12.159y + 8.312z = 8.777,$$
$$3.328x - 6.172y + 2.455z = 0.233.$$

4. Solve
$$3.78x - 2.36y + 6.16z = 18.72,$$
$$.95x - 17.15y - 11.33z = 8.69,$$
$$2.87x + 4.14y - 7.81z = -3.90.$$

3. THE METHOD OF RELAXATION

The method of relaxation differs from the two methods just discussed in that it obtains a solution for all unknowns simultaneously. This solution is an approximation to a certain number of decimals; at each step in the solution, an additional decimal place in each unknown is obtained. When only two or three decimal places are required, as is the case in many engineering problems, the method of relaxation avoids carrying a large number of figures throughout the work.

The basic principle of the method is as follows. Let the equations be

$$\sum_{j=1}^{n} a_{ij}x_j - c_i = 0 \quad (i = 1, 2, \dots, n).$$

Set

$$R_i = \sum_{j=1}^{n} a_{ij}x_j - c_i.$$

Then the solution of the n equations is a set of numbers $x_1, x_2, \dots, x_n$, having the property that it makes all the R_i equal to zero. We shall obtain an approximate solution by employing an iterative procedure which makes the R_i smaller and smaller at each step (thus we shall get closer and closer to the actual solution). The quantities R_i are known as the *residuals* of the n equations. The actual procedure can best be outlined in an example.

EXAMPLE 8.4.

$$3x + 9y - 2z - 11 = 0,$$
$$4x + 2y + 13z - 24 = 0,$$
$$11x - 4y + 3z + 8 = 0.$$

We start with the following *operations table*.

Δx	Δy	Δz	ΔR_1	ΔR_2	ΔR_3
1	0	0	3	4	11
0	1	0	9	2	-4
0	0	1	-2	13	3

Entries in this table are interpreted, for example, as "An increment of 1 unit in x produces an increment of 3 units in R_1, 4 units in R_2, and 11 units in R_3".

We now start the solution by taking any trial solution ($x = y = z = 0$ is usually simplest), computing the corresponding values of R_i, and then using the operations table to change the x, y, and z so as to decrease the R_i. The first stage of the solution then follows.

Trial solutions appear between the two horizontal bars; the other lines show the changes or *relaxations* in the values of the unknowns, together with the resulting cumulative totals for the residuals.

$x=0$	$y=0$	$z=0$	$R_1 = -11$	$R_2 = -24$	$R_3 = 8$
0	0	2	-15	2	14
-1	0	0	-18	-2	3
0	2	0	0	2	-5
-1	2	2	0	2	-5

Our second trial solution is $x = -1$, $y = 2$, $z = 2$, making $R_1 = 0$, $R_2 = 2$, $R_3 = -5$. Rather than introduce decimal fractions of an increment, we multiply through by 10 and continue the procedure.

-10	20	20	0	20	-50
5	0	0	15	40	5
0	0	-3	21	1	-4
0	-2	0	3	-3	4
-5	18	17	3	-3	4

Our third trial solution is $10x = -5$, $10y = 18$, $10z = 17$. To obtain the second decimal place in the solution, we again multiply by 10 and relax.

−50	180	170		30	−30	40
−4	0	0		−18	−46	−4
0	0	4		−10*	6	8
0	1	0		−1	8	4
0	0	−1		1	−5	1
−54	181	173		1	−5	1

Let us now check by substituting these values $100x = -54$, $100y = 181$, $100z = 173$. We obtain residuals 21, −5, 1.

−54	181	173		21	−5	1
0	−2	0		3	−9	9
−1	0	0		0	−13	−2
0	0	1		−2	0	1
−55	179	174		−2	0	1

We had made an error (actually in the entry marked with an asterisk). Note, however, that we do *not* need to retrace our work and find the error. We simply start from the correct residuals and relax until they are small.

Completing the solution to five decimals gives

−550	1790	1740		−20	0	10
0	2	0		−2	4	2
−5500	17920	17400		−20	40	20
0	2	0		−2	44	12
0	0	−3		4	5	3
−55000	179220	173970		40	50	30
0	−4	0		4	42	46
−4	0	0		−8	26	2
0	0	−2		−4	0	−4
−55004	179216	173968		−4	0	−4

We thus conclude that $10^5x = -55004$, $10^5y = 179216$, $10^5z = 173968$, and that these values make $R_1 = -.00004$, $R_2 = 0$, $R_3 = -.00004$. These are checked by substitution.

Various subtleties in the relaxation procedure can only be learned by practice; unlike many simple numerical procedures, it assumes skill on the part of the person using it, and so is not well suited for programming for an electronic computer. We might note several points about relaxation.

(a) It is advantageous if exactly one large coefficient appears in each column of the operations table; in this way, one can eat away at one residual without greatly altering the others.

(b) It is generally best to attack first those residuals which, in changing, have the most effect on the other residuals. For example, in the table, we would tend to cut down R_1 and R_3 first, since changes in both of these residuals produce greater relative over-all changes than does a change in R_2, where the increment -2, 13, 3, is mainly concentrated in R_2.

(c) Relaxation is an art; no two persons will practise it in exactly the same way, and no general rules of procedure can be given. Some students like to think that the rule "Always liquidate the largest residual" is valid. But this attempted rule must often yield to the consideration in (b). Indeed, it may often be well to under-relax (that is, not liquidate a residual as much as one might) or over-relax (that is, more than wipe out a residual, so that it remains appreciable with its sign altered) if foresight indicates that such a procedure will make further relaxations more effective. In this respect, relaxation resembles a chess game, and this is one of the great values of the method; it replaces the drudgery of slugging out a solution by the mental stimulation of a game. You are fighting the equations, and you are attempting to select the best strategy available. Perhaps this is the closest approach to a general rule one can make—if a step is opportune, that is, if it advances you nearer to winning the battle of decreasing some residuals without unduly exposing you to a flank attack (namely, a more unmanageable increase in other residuals), then it is a good step.

We conclude by giving a solution as it might normally be presented in tabular form.

EXAMPLE 8.5.

$$2x_1 - 13x_2 - 3x_3 - 49 = 0,$$

$$5x_1 - 6x_2 + 17x_3 - 25 = 0,$$

$$11x_1 + 2x_2 - 4x_3 + 31 = 0.$$

Δx_1	Δx_2	Δx_3	ΔR_1	ΔR_2	ΔR_3
1	0	0	2	5	11
0	1	0	−13	−6	2
0	0	1	−3	17	−4
$x_1 = 0$	$x_2 = 0$	$x_3 = 0$	$R_1 = -49$	$R_2 = -25$	$R_3 = 31$
	−4		3	−1	23
−2			−1	−11	1
		1	−4	6	−3
−20	−40	10	−40	60	−30
	−3		−1	78	−36
3			5	93	−3
		−5	20	8	17
	1		7	2	19
−2			3	−8	−3
−190	−420	50	30	−80	30
		5	15	5	−50
5			25	30	5
	2		−1	18	9
		−1	2	1	13
−1			0	−4	2
−1860	−4180	540	0	−40	20
−2			−4	−50	−2
		3	−13	1	−14
	−1		0	7	−16
1			2	12	−5
		−1	5	−5	1
−1861	−4181	542	5	−5	1

We conclude that $10^3x = -1861$, $10^3y = -4181$, $10^3z = 542$. Continued relaxation will produce the answer to any required degree of accuracy.

EXERCISES

1. Continue Example 8.5 to give $x = -1.860827$, $y = -4.180674$, $z = .542361$ with residuals 5×10^{-6}, -4×10^{-6}, 10^{-6}.

2.

$$11x - 2y + 3z - 37 = 0,$$
$$2x + 3y + 13z + 41 = 0,$$
$$5x - 18y + 7z - 23 = 0.$$

Obtain the solution up to the stage $x = 4.02745$, $y = 1.49255$, $z = -3.42902$ with residuals -10^{-5}, -10^{-5}, $+10^{-5}$.

3. Solve

(a) $4x + 17y + 92 = 0,$
      $15x - 7y - 87 = 0.$

(b) $2x - 3y + 8z = 93,$
      $x + 5y - 2z = 119,$
      $2x + 7y + 15z = 137.$

(c) $2x - 12y + 5z = 37,$
      $13x + 3y + 2z = 115,$
      $3x - 5y + 15z = -86.$

4. Solve the equations

$$2x + 3y + 4z + 14u = 87,$$
$$12x + 6y + 5z - 3u = 72,$$
$$3x + 17y + 8z + 5u = 105,$$
$$2x + 3y + 15z + 2u = 92.$$

5. Carry the solution of the equations

$$8x - 3y + z - 37 = 0,$$
$$x + 11y - 5z - 12 = 0,$$
$$3x + 2y - 19z + 17 = 0.$$

far enough to give $x = 4.9486$, $y = 1.4734$, $z = 1.8312$.

6. Use the method of relaxation to obtain the solution to four decimals of the equations

$$3x - 14y + 2z + 15 = 0,$$
$$17x + 5y - 3z - 37 = 0,$$
$$4x - 3y + 13z + 53 = 0.$$

4. GROUP RELAXATION

The method of relaxation is not immediately suitable if the coefficients are roughly equal. For, in such a case, any attempt to change one residual will produce a roughly equal alteration in the other residuals. Whittling away at the residuals may then become rather slow and unsatisfactory. So we attempt to construct an operations table where combinations of increments are chosen so as to give us the sort of situation we want, namely, one in which each operation has its major effect on one residual only.

EXAMPLE 8.6.

$$10x + 9y - 13z - 71 = 0,$$
$$8x - 7y - 14z + 35 = 0,$$
$$9x + 10y + 12z - 152 = 0.$$

Our initial operations table is

1	0	0	10	8	9
0	1	0	9	−7	10
0	0	1	−13	−14	12

This is clearly unsatisfactory. However, by very simple combinations, we get

1	−1	0		1	15	−1
1	0	1		−3	−6	21
0	1	−1		22	7	−2

Sometimes slightly more complicated combinations will be needed to produce the desired pattern. The solution now proceeds as before.

0	0	0		−71	35	−152

7		7		−92	−7	−5
	4	−4		−4	21	−13
−1	1			−5	6	−12

60	50	30		−50	60	−120

6		6		−68	24	6
	3	−3		−2	45	0
−3	3			−5	0	3

630	560	330		−50	0	30

−1		−1		−47	6	9
	2	−2		−3	20	5
−1	1			−4	5	6

628	563	327		−4	5	6

We thus have $x = 6.28$, $y = 5.63$, $z = 3.27$. This solution checks.

EXERCISES

1. Continue Example 8.6 to give the solution to four decimals.

2. Solve to three places of decimals.

   (a)
$$13x - 9y = 116,$$
$$15x + 11y = 82.$$

   (b)
$$17x + 11y - 6z = 132,$$
$$18x - 12y + 8z = 103,$$
$$15x - 13y + 7z = 95.$$

   (c)
$$5x + 17y - 12z + 10u - 86 = 0,$$
$$6x + 13y + 15z + 12u - 103 = 0,$$
$$4x + 20y - 13z + 9u - 94 = 0,$$
$$8x - 18y + 2z + 3u - 7 = 0.$$

3. Solve to four places of decimals the equations

$$6x - 5y + 8z = 41,$$
$$3x - 4y - 2z = 55,$$
$$7x + 7y - 5z = 43.$$

5. EQUATIONS WITH UNWIELDY COEFFICIENTS

Most sets of equations actually encountered in practice have coefficients which are not simple small integers, but rather cumbersome decimals. One method of procedure is exemplified in

EXAMPLE 8.7.

$$1.637x + 2.349y - 8.371z - 3.497u = 6.354,$$
$$2.398x + 1.763y + 1.118z + .897u = 2.953,$$
$$4.444x - 4.986y - 7.677z - 3.400u = 1.700,$$
$$2.009x + 4.906y - 8.089z + 4.098u = -3.870.$$

When one is first confronted with a set of equations such as the above, one is generally taken aback. Indeed, I have often tried such a problem on unsuspecting students, only to find them, an hour later, still sitting looking at the equations, with no idea how to proceed. Yet such a problem is both common and important.

To begin, one may decide to apply the method of Section 2; however, the relaxation process can be adapted so as to give a solution with less painful a struggle than one might anticipate. We first multiply the equations by 10 and round to integral values, thus obtaining

$$16x + 23y - 84z - 35u - 64 = 0,$$
$$24x + 18y + 11z + 9u - 30 = 0,$$
$$44x - 50y - 77z - 34u - 17 = 0,$$
$$20x + 49y - 81z + 41u + 39 = 0.$$

There are two reasons for this step; first, we want to get an approximate solution so that the first trial solution in our final relaxation will be good; secondly, with smaller numbers we more readily decide on the appropriate compound relaxations. Our operations table becomes

1	0	0	0	16	24	44	20
0	1	0	0	23	18	−50	49
0	0	1	0	−84	11	−77	−81
0	0	0	1	−35	9	−34	41
0	0	1	−2	−14	−7	−9	−163
1	−1	0	0	−7	6	94	−29
0	1	0	−1	58	9	−16	8
3	1	1	0	−13	101	5	28

Using the group relaxation table gives

0	0	0	0	−64	−30	−17	39
0	1	0	−1	−6	−21	−33	47
0	10	0	−10	−60	−210	−330	470
6	2	2	0	−86	−8	−320	526
3	−3	0	0	−107	10	−38	439
0	2	0	−2	9	28	−70	455
0	0	3	−6	−33	7	−97	−34
1	−1	0	0	−40	13	−3	−63
0	1	0	−1	18	22	−19	−55

So we have $x = 1.0, y = 1.1, z = .5, u = -1.9$. This is a very rough solution, but we can now use it on the original equations. The operations table then becomes (multiplying all original equations by 1000)

0	0	1	−2	−1377	−676	−877	−16285
1	−1	0	0	−712	635	9430	−2897
0	1	0	−1	5846	866	−1586	808
3	1	1	0	−1111	10075	669	2844

We then set up the relaxation table, starting with the value $10(x, y, z, u) = (10, 11, 5, -19)$, and relax further.

10	11	5	−19	3257	2390	−1191	−5551
100	110	50	−190	32570	23900	−11910	−55510
0	0	−3	6	36701	25928	−9279	−6655
0	−6	0	6	1625	20732	237	−11503
−6	−2	−2	0	3847	582	−1101	−17191
0	0	−1	2	5224	1258	−224	−906
0	−1	0	1	−622	392	1362	−1714
940	1010	440	−1750	−6220	3920	13620	−17140
−1	1	0	0	−5508	3285	4190	−14243
0	1	0	−1	338	4151	2604	−13435
0	0	−1	2	1715	4827	3481	2850
9390	10120	4390	−17490	17150	48270	34810	28500
−4	4	0	0	19998	45730	−2910	40088
0	−3	0	3	2460	43132	1848	37664
−12	−4	−4	0	6904	2832	−828	26288
0	0	2	−4	4150	1480	−2582	−6282
0	−1	0	1	−1696	614	−996	−7090

Adding, our solution at this stage is

9376	10116	4388	−17490	−1696	614	−996	−7090

Let us now check by computing residuals from the actual equations; we get

9376	10116	4388	−17490	1578	5410	7892	−3072

Notice that every residual was incorrect! However, *we need not correct the errors, but simply start from this set of correct residuals.* A single relaxation produces

9375	10117	4388	−17490	2290	4775	−1538	−175

So we conclude that $x = .9375$, $y = 1.0117$, $z = .4388$, $u = −1.7490$, the final digits being dubious.

EXERCISES

1. In Example 8.7, continue the approximate solution to the stage $100(x, y, z, u)$ before starting on the final relaxation table.

2. In Example 8.7, continue the final solution to another place of decimals.

3. Solve, by relaxation, the systems of equations

   (a) $$3.271x + 5.924y = 6.295,$$
   $$5.417x − 6.231y = 2.444.$$

   (b) $$3.195x − 2.764y + 8.542z = 7.245,$$
   $$5.006x + 3.914y − 7.887z = 1.024,$$
   $$4.497x + 2.250y + 6.338z = 14.118.$$

   (c) $$4.73x − 7.21y + 6.52z + 1.98u − 13.87 = 0,$$
   $$8.62x + 5.91y − 3.76z − 8.12u + 19.51 = 0,$$
   $$8.17x − 5.23y − 9.92z − 7.67u + 8.69 = 0,$$
   $$3.67x − 9.21y − 2.67z + 12.34u + 3.69 = 0,$$

   (d) $$3.142x + 2.183y + .982z + 1.764u = 10.211,$$
   $$4.968x − 4.311y + 1.725z − 2.900u = 3.427,$$
   $$5.194x − 2.777y − 2.134z + 2.455u = 3.313,$$
   $$4.068x + 3.667y − 2.195z + 4.373u = 2.112.$$

6. ITERATIVE METHODS OF SOLUTION

The relaxation method starts from a guessed solution and inches gradually up towards a true solution. However, it is not an iterative procedure, in that no definite routine defines the way in which one should proceed from one approximate solution to the next approximate solution. We shall now work two examples illustrating two types of iterative process, and then discuss them more generally.

EXAMPLE 8.8.

$$13x_1 + 5x_2 - 3x_3 + x_4 = 18,$$
$$2x_1 + 12x_2 + x_3 - 4x_4 = 13,$$
$$3x_1 - 4x_2 + 10x_3 + x_4 = 29,$$
$$2x_1 + x_2 - 3x_3 + 9x_4 = 31.$$

The first step is to use the successive equations to "solve" for each unknown in terms of the others. Thus we obtain

$$x_1 = \tfrac{1}{13}(18 - 5x_2 + 3x_3 - x_4),$$
$$x_2 = \tfrac{1}{12}(13 - 2x_1 - x_3 + 4x_4),$$
$$x_3 = \tfrac{1}{10}(29 - 3x_1 + 4x_2 - x_4),$$
$$x_4 = \tfrac{1}{9}(31 - 2x_1 - x_2 + 3x_3).$$

Let us start with the trial solution $x_1 = x_2 = x_3 = x_4 = 0$ and substitute in the right-hand side of this system of equations (working to three decimals). If our trial solution works, then the left-hand side should be the same. Actually, it gives

$$x_1 = 1.385, \quad x_2 = 1.083, \quad x_3 = 2.900, \quad x_4 = 3.444.$$

This set of values then becomes our second trial solution. Substituting it on the right-hand side of our system of equations, we obtain

$$x_1 = 1.372, \quad x_2 = 1.759, \quad x_3 = 2.573, \quad x_4 = 3.983.$$

Continue this procedure; at each stage, the answer obtained is used as the next trial solution for substitution. The remainder of the work can be arranged as

.995	.980	1.025	1.005	.996	1.001	1.001
1.968	1.952	1.981	2.001	1.999	1.998	2.000
2.794	3.009	2.993	2.984	3.000	3.002	2.999
3.802	3.936	4.013	3.994	3.993	4.001	4.001

Since it is easily checked that the exact answer is

$$x_1 = 1, \quad x_2 = 2, \quad x_3 = 3, \quad x_4 = 4,$$

we see that the iteration has finally converged to the solution. While no simple necessary and sufficient conditions for convergence are known, it is clear that if the procedure does converge, it automatically produces a set of values which satisfies both the left-hand and right-hand sides of the equations simultaneously, that is, it produces the solution.

The procedure of Example 8.8 is known as the Jacobi method of iteration. We shall now illustrate an alternative method.

EXAMPLE 8.9. Take the equations of Example 8.8 in the form

$$x_1 = \tfrac{1}{13}(18 - 5x_2 + 3x_3 - x_4),$$
$$x_2 = \tfrac{1}{12}(13 - 2x_1 - x_3 + 4x_4),$$
$$x_3 = \tfrac{1}{10}(29 - 3x_1 + 4x_2 - x_4),$$
$$x_4 = \tfrac{1}{9}(31 - 2x_1 - x_2 + 3x_3).$$

We again start with the trial solution $x_1 = x_2 = x_3 = x_4 = 0$. However, this time we substitute only in the first equation to give $x_1 = 1.385$. We then substitute $x_1 = 1.385$, $x_2 = x_3 = x_4 = 0$ in the second equation to obtain $x_2 = .853$. Put $x_1 = 1.385$, $x_2 = .853$, $x_3 = x_4 = 0$ in the third equation to give $x_3 = 2.826$. Finally, putting $x_1 = 1.385$, $x_2 = .853$, $x_3 = 2.826$, $x_4 = 0$, in the fourth equation yields $x_4 = 3.984$.

We see that in this method of iteration, known as the Gauss-Seidel method, the result of any stage within a step is used in succeeding stages of the same step. The remaining steps of the iteration, proceeding from the second trial solution, follow.

1.385	1.402	1.000	1.012	1.000
.853	1.942	1.969	1.999	1.999
2.826	2.858	3.001	2.996	3.000
3.984	3.870	4.004	3.996	4.000

The same remark applies to the Gauss-Seidel method as to the Jacobi method: no simple necessary and sufficient conditions for convergence are known; however, if the process does converge, it automatically produces the solution.

We must now consider how best to adapt an arbitrary system of equations to the Jacobi or Gauss-Seidel methods. The examples show it is advantageous to arrange the unknowns in such an order that the diagonal terms are large; then each unknown can be expressed in a form involving a sizable numerical denominator. A method of arrangement can easily be illustrated by

EXAMPLE 8.10. Arrange the following system for iteration:

$$3x_1 - 5x_2 + 47x_3 + 20x_4 = 18,$$
$$56x_1 + 23x_2 + 11x_3 - 19x_4 = 36,$$
$$12x_1 + 16x_2 + 17x_3 + 18x_4 = 25,$$
$$17x_1 + 65x_2 - 13x_3 + 7x_4 = 84.$$

We begin by writing down the matrix of coefficients in the form

$$
\begin{array}{rrrr}
3 & -5 & 47 & 20 \\
56 & 23 & 11 & -19 \\
12 & 16 & 17 & 18 \\
17 & 65 & -13 & 7
\end{array}
$$

We must now select four large matric elements with the property that only one is in each row and only one is in each column. If we proceed by always choosing the maximal entry, we start with 65 and, deleting to find the corresponding minor, obtain

$$
\begin{array}{rrr}
3 & 47 & 20 \\
56 & 11 & -19. \\
12 & 17 & 18
\end{array}
$$

We then choose 56, 47, and 18, in that order; so our pivotal coefficients become 65, 56, 47, 18. While this procedure is a definite one, there may be other selections which are better in any particular problem; in this problem, 65, 56, 17, 20, is an alternative choice. Whether it would work better can only be determined by trial.

With the choice of 65, 56, 47, 18, as pivotal coefficients, the equations can be written as

$$
\begin{aligned}
\underline{65x_2} + 17x_1 - 13x_3 + 7x_4 &= 84, \\
23x_2 + \underline{56x_1} + 11x_3 - 19x_4 &= 36, \\
-5x_2 + 3x_1 + \underline{47x_3} + 20x_4 &= 18, \\
16x_2 + 12x_1 + \overline{17x_3} + \underline{18x_4} &= 25.
\end{aligned}
$$

The iteration scheme then sets

$$
\begin{aligned}
x_2 &= \tfrac{1}{65}(84 - 17x_1 + 13x_3 - 7x_4), \\
x_1 &= \tfrac{1}{56}(36 - 23x_2 - 11x_3 + 19x_4), \\
x_3 &= \tfrac{1}{47}(18 + 5x_2 - 3x_1 - 20x_4), \\
x_4 &= \tfrac{1}{18}(25 - 16x_2 - 12x_1 - 17x_3).
\end{aligned}
$$

We are now prepared for

EXAMPLE 8.11. Solve Example 8.10 by the Jacobi iteration technique.

The first twelve iterations are

$$
\begin{array}{rrrrrrr}
1.29 & 1.05 & 1.20 & 1.39 & 1.39 & 1.49 & 1.51 \\
.64 & .51 & .05 & .09 & -.13 & -.16 & -.23 \\
.38 & -.11 & .69 & .41 & .68 & .66 & .72 \\
1.39 & -.54 & .22 & -.36 & -.29 & -.40 & -.45
\end{array}
$$

$$
\begin{array}{rrrrr}
1.54 & 1.56 & 1.5797 & 1.5901 & 1.5982 \\
-.27 & -.30 & -.3221 & -.3382 & -.3503 \\
.75 & .77 & .7851 & .7950 & .8044 \\
-.48 & -.51 & -.5250 & -.5420 & -.5498
\end{array}
$$

At this stage, we can cease checking our calculations, since minor errors are unimportant; since we always return to the original set of equations, the Jacobi process is self-correcting. Any major error would make a large change in some x-value, and so could not be missed. The next iterations produce

1.6040	1.6079	1.6109	1.6129	1.6144	1.6155
−.3581	−.3642	−.3683	−.3713	−.3734	−.3749
.8093	.8139	.8166	.8188	.8202	.8214
−.5579	−.5625	−.5662	−.5687	−.5706	−.5719

We note that convergence is quite slow; however, we can help it out by looking at the differences of this last set of iterations. These are

39	30	20	15	11
−61	−41	−30	−21	−15
46	27	22	15	11
−46	−37	−25	−19	−13

These differences appear to approximate a geometric progression with common ratio $\frac{3}{4}$. If we assume that the next terms are 8, −11, 8, −10, respectively, and sum all succeeding differences to infinity, we obtain the sums to infinity as 32, −44, 32, −40. Using these as corrections for a new start in the iteration, we find (note that the underlined minor error in addition is immaterial)

1.6187	1.6185	1.6184	1.6184	1.6183
−.3794	−.3793	−.3792	−.3791	−.3790
.8246	.8245	.8244	.8243	.8242
−.5759	−.5758	−.5756	−.5755	−.5755

The final steps in the iteration merely involve inching back and forth with a little give-and-take among the various values.

1.6182	1.6182	1.6182
−.3790	−.3788	−.3788
.8242	.8241	.8241
−.5753	−.5753	−.5753

We thus conclude that the answer is

$$x_2 = 1.6182, \quad x_1 = -.3788, \quad x_3 = .8241, \quad x_4 = -.5753.$$

There is, conceivably, an error of as much as one unit in some of the final digits.

Obviously, the Jacobi process may, when convergence is slow, as in this example, become quite laborious. However, since it follows a perfectly well-defined routine, and since one merely inserts the result of any step in the original system of equations in order to proceed with the next step, the method is a good one for an electronic computer. One might say that the method follows a definite "programme".

EXAMPLE 8.12. Apply the Gauss-Seidel iteration technique to the equations of Example 8.11.

We follow the same pattern as when using Jacobi iteration, and obtain

1.29	1.40	1.51	1.55	1.58
.11	−.14	−.24	−.30	−.33
.51	.67	.73	.77	.79
−.31	−.40	−.48	−.52	−.54

At this stage, we discontinue checking, since convergence allows us to detect any large error. The next iterations produce

1.5948	1.6047	1.6104	1.6137	1.6156
−.3505	−.3626	−.3695	−.3734	−.3757
.8048	.8130	.8177	.8204	.8220
−.5551	−.5636	−.5685	−.5714	−.5731

Taking differences, we find

99	57	33	18
−121	−69	−39	−23
72	47	27	16
−85	−49	−29	−17

These values suggest a common ratio of .6; using this value, with initial differences of 11, −14, 10, −10, yields correction terms of 28, −35, 25, −25 (again found by taking the sums to infinity of the geometric progressions of differences). Finishing the iteration, we have

1.61840	1.61837	1.61828	1.61821	1.61820
−.37920	−.37908	−.37894	−.37887	−.37884
.82450	.82428	.82419	.82415	.82412
−.57560	−.57543	−.57536	−.57530	−.57529

1.61817	1.61816	1.61815	1.61814	1.61814
−.37882	−.37880	−.37879	−.37879	−.37878
.82411	.82409	.82409	.82409	.82408
−.57526	−.57525	−.57525	−.57524	−.57524

One further iteration merely reproduces the last set of values. So we conclude that

$$x_2 = 1.61814, \quad x_1 = -.37878, \quad x_3 = .82408, \quad x_4 = -.57524.$$

We can repeat here our remarks concerning the Jacobi method, namely, that the last digits may be in error by 1 unit, and that the method is well-suited for programming on an electronic computer.

EXERCISES

1. The solution of
$$7x - 2y = 11, \qquad 3x + 13y = 10,$$
is readily found to be $x = 1.680412$, $y = .381443$. Arrange these two equations in a form suitable for iteration, and, starting from $(0, 0)$, apply nine Jacobi iterations to give the solution (work to four decimals). Thence write down the sequence of five Gauss-Seidel iterations which produces the answer.

2. Find the solution, to four decimals, of
$$\begin{aligned} 83x_1 + 11x_2 - 4x_3 &= 95, \\ 7x_1 + 52x_2 + 13x_3 &= 104, \\ 3x_1 + 8x_2 + 29x_3 &= 71. \end{aligned}$$
Use both the Jacobi and Gauss-Seidel techniques.

3. Apply (carrying four decimals) the Jacobi and Gauss-Seidel techniques to

   (a)
$$\begin{aligned} 3x_1 + 7x_2 + 36x_3 - 2x_4 &= 18, \\ 17x_1 + 2x_2 - 5x_3 + 3x_4 &= 19, \\ 4x_1 - 43x_2 + 11x_3 - 7x_4 &= 56, \\ 5x_1 - 8x_2 - 3x_3 + 38x_4 &= -15. \end{aligned}$$

   (b)
$$\begin{aligned} 2x_1 + 3x_2 + 4x_3 + 3x_4 + 27x_5 &= 19, \\ x_1 - 2x_2 - 2x_3 + 85x_4 - 6x_5 &= 28, \\ 3x_1 - 67x_2 + 15x_3 - 7x_4 + 5x_5 &= 39, \\ 53x_1 + 5x_2 - 6x_3 - 4x_4 - 3x_5 &= -112, \\ 6x_1 - 6x_2 + 75x_3 + 19x_4 + 2x_5 &= 24. \end{aligned}$$

4. Apply (carrying five decimals) the Jacobi and Gauss-Seidel techniques to
$$\begin{aligned} 17x + 294y - 35z &= 812, \\ 513x - 35y + 42z &= 1272, \\ 29x + 36y + 312z &= 519. \end{aligned}$$

5. Find, by iteration, the largest eigenvalue of the matrix of coefficients
$$\begin{pmatrix} 17 & 294 & -35 \\ 513 & -35 & 42 \\ 29 & 36 & 312 \end{pmatrix}$$
occurring in Exercise 4.

6. Apply the Jacobi iteration procedure to the equations
$$\begin{aligned} 17a + 65b - 13c + 50d &= 84, \\ 12a + 16b + 37c + 18d &= 25, \\ 56a + 23b + 11c - 19d &= 36, \\ 3a - 5b + 47c + 10d &= 18. \end{aligned}$$
Carry two decimals and proceed to twelve or more iterations.

7. Apply the Gauss-Seidel procedure in Exercise 6.

7. ILL-CONDITIONED EQUATIONS

Let us begin with

EXAMPLE 8.13.

$$x + 5y = 17,$$

$$1.500x + 7.501y = 25.503.$$

The solution is readily found to be $x = 2$, $y = 3$.
On the other hand, the solution of

$$x + 5y = 17,$$

$$1.500x + 7.501y = 25.500,$$

is $x = 17$, $y = 0$.

Clearly, these equations behave very badly; a change of one part in 8000 in one of the coefficients is sufficient to change the solution from (2, 3) to (17, 0). Such systems of equations, where the solution is very sensitive to tiny alterations in the coefficients, are called *ill-conditioned*. In fact, if the coefficients of ill-conditioned equations are the result of an experiment, then no information at all is, in general, available about the solution. To all intents and purposes, the two equations in this example are (experimentally) one and the same equation. Of course, if the coefficients are not obtained experimentally, then the two equations indeed differ, and a unique solution exists. However, obtaining the solution may occasion considerable difficulty. For instance, in the equations

$$x + 5y - 17 = 0,$$

$$1.5x + 7.501y - 25.503 = 0,$$

we find $x = 17$, $y = 0$, makes $R_1 = 0$, $R_2 = -.002$. Certainly these residuals are small, yet the true solution is (2, 3). This illustrates a fundamental fact about ill-conditioned equations: *there are values, in no way resembling the true solution, which make the residuals very small.*

This fact is even more strikingly pointed up in

EXAMPLE 8.14. [J. Morris, *Phil. Mag.* **7**, 37 (1946), 106].

$$5x + 7y + 6z + 5u = 23,$$

$$7x + 10y + 8z + 7u = 32,$$

$$6x + 8y + 10z + 9u = 33,$$

$$5x + 7y + 9z + 10u = 31.$$

Obviously, the solution is $x = y = z = u = 1$. However, the values $x = 2.36$, $y = .18$, $z = .65$, $u = 1.21$, make $R_1 = .01$, $R_2 = -.01$, $R_3 = -.01$, $R_4 = .01$. These equations are very ill-conditioned.

By and large, if the equations are ill-conditioned, only the method of elimination given in Section 1 is a foolproof method of solution. The round-off error in triangulation may be far too great if the solution is sensitive, and we have just seen that relaxation may fail, since the residuals may be small without our having approached closely to a solution.

EXERCISES

1. In Example 8.14, find the residuals for

$$x = 14.6, \quad y = -7.2, \quad z = -2.5, \quad u = 3.1.$$

8. A MEASURE OF ILL-CONDITION

Let

$$\sum_{j=1}^{n} a_{ij}x_j - c_i = 0 \quad (i = 1, 2, \ldots, n)$$

be the given set of n equations. Clearly the solutions $x_j = D_j/D$ will tend to fluctuate badly if and only if $D = \det(a_{ij})$ is small in relation to D_j, or, one might say, small in relation to the coefficients in the set of equations. However, since evaluating $\det(a_{ij})$ is not an exceptionally pleasant prospect, this criterion concerning the size of D is not of much use.

We can get a good picture of the geometric significance of ill-condition for the case of two equations. Let

$$R_1 = a_{11}x + a_{12}y - c_1,$$
$$R_2 = a_{21}x + a_{22}y - c_2.$$

Then the equations $R_1 = 0$, $R_2 = 0$, represent a pair of straight lines (Figure 8.1) intersecting at point P.

Suppose, however, we do not demand that $R_1 = R_2 = 0$, but only that $|R_1| < \epsilon_1$, $|R_2| < \epsilon_2$, where ϵ_1 and ϵ_2 are small positive numbers. Then we only demand that the "solution" lie in a narrow band (bounded by the lines $R_1 \pm \epsilon_1 = 0$) on either side of $R_1 = 0$ and in another narrow band (bounded by the lines $R_2 \pm \epsilon_2 = 0$) on either side of $R_2 = 0$. If the two lines meet at a moderately sharp angle, this gives us an approximate solution, since the coordinates of any point in the shaded parallelogram will be near to those of the true solution P.

If, on the other hand, the two lines in Figure 8.1 are nearly coincident, then the shaded parallelogram will be extremely long and narrow, and there will be points within it, situated a very great distance from P, which make $|R_1| < \epsilon_1$ and $|R_2| < \epsilon_2$.

A generalization of this concept provides us with one possible measure of ill-condition. If $n = 3$, the equations

$$\sum_{j=1}^{3} a_{ij}x_j - c_i = 0 \quad (i = 1, 2, 3)$$

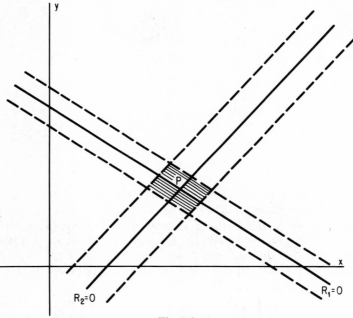

Fig. 8.1.

represent three planes. The angle between any two of them, say the first and second, is given by

$$\cos \theta_{12} = \frac{a_{11}a_{21} + a_{12}a_{22} + a_{13}a_{23}}{\sqrt{\Sigma a_{1j}^2}\sqrt{\Sigma a_{2j}^2}}.$$

Generalizing this formula to n dimensions, we say that n linear equations represent n hyperplanes and that the $\binom{n}{2}$ angles between them are given by

$$\cos \theta_{ij} = \frac{\displaystyle\sum_{k=1}^{n} a_{ik}a_{jk}}{\sqrt{\displaystyle\sum_{k=1}^{n} a_{ik}^2}\sqrt{\displaystyle\sum_{k=1}^{n} a_{jk}^2}}.$$

If certain of the θ_{ij} are very small, that is, the cosines are nearly 1, then the equations are ill-conditioned.

EXAMPLE 8.15. In Example 8.14, we have

$$\cos^2 \theta_{12} = \frac{188^2}{(135)(262)} = \frac{35344}{35370} = .999,$$

$$\cos^2 \theta_{13} = \frac{191^2}{(135)(281)} = \frac{36481}{37935} = .962,$$

$$\cos^2 \theta_{14} = \frac{178^2}{(135)(255)} = \frac{31684}{34425} = .920,$$

$$\cos^2 \theta_{23} = \frac{265^2}{(262)(281)} = \frac{70225}{73622} = .954,$$

$$\cos^2 \theta_{24} = \frac{247^2}{(262)(255)} = \frac{61009}{66810} = .913,$$

$$\cos^2 \theta_{34} = \frac{266^2}{(281)(255)} = \frac{70756}{71655} = .987.$$

EXAMPLE 8.16. Consider Example 8.4, where we found the equations well-conditioned. There

$$\cos^2 \theta_{12} = \frac{4^2}{(94)(189)} = .001,$$

$$\cos^2 \theta_{13} = \frac{(-9)^2}{(94)(146)} = .006,$$

$$\cos^2 \theta_{23} = \frac{75^2}{(189)(146)} = .204.$$

The question next arises as to how small θ should be (or how close $\cos^2 \theta$ should be to 1) before we label the equations "ill-conditioned". Inevitably, a certain degree of arbitrariness is involved here. If we consider the shaded parallelogram in Figure 8.1, and select ϵ_1 and ϵ_2 so that it is a rhombus, then the diagonals are equal to $4a \cos \frac{\theta}{2}$ and $4a \sin \frac{\theta}{2}$, where $2a$ is the side of the rhombus and θ is the angle at which the two lines meet. The ratio of the two diagonals is $\cot \frac{\theta}{2}$. If we arbitrarily decide that the dividing line between well-conditioned and ill-conditioned equations occurs when this ratio passes 5, we find that $\theta = 22°37'$, $\cos^2 \theta = .9231^2 = .852$. On the other hand, if we set the dividing line at the ratio 10 to 1, we find $\theta = 11°24'$, $\cos^2 \theta = .961$. Wherever we agree to set the line, the distinction between the situations represented in Examples 8.15 and 8.16 is obvious. (Quite clearly, we are only using the two-dimensional case, illustrated in Figure 8.1, as a guide; we might prefer to decide arbitrarily to set the dividing line at $\cos^2 \theta = .90$ or $\cos^2 \theta = .95$ without any argument by analogy.)

EXERCISES

1. Prove that the diagonals of the parallelogram in Figure 8.1 are indeed $4a \cos \dfrac{\theta}{2}$ and $4a \sin \dfrac{\theta}{2}$ when ϵ_1 and ϵ_2 are selected so that the figure is a rhombus.

2. Compute $\cos^2 \theta_{ij}$ for
$$3x + 7y - 5z = 10,$$
$$2x - 4y + 9z = 8,$$
$$5x + 6y + 2z = 12.$$

3. Compute $\cos^2 \theta_{ij}$ for the equations given in Exercises 3 and 4 of Section 3.

4. Show that the condition developed for ill-condition, though sufficient, is not necessary by considering the following $n \times n$ matrix, communicated by W. M. Kahan:
$$a_{ii} = [(n - 1)/n]^{1/2}, \qquad a_{ij} = -[n(n - 1)]^{-1/2} \quad \text{for} \quad i \neq j$$

Prove: (a) A is a singular matrix;

(b) $\cos \theta_{ij} = -1/(n - 1)$ which is small for n large.

CHAPTER NINE

Difference Equations

1. INTEGRATION AND SUMMATION

In a direct integration problem, that is, a differential equation in its simplest form, we are given dy/dx and required to find y. Thus, if $dy/dx = 2x$, we write

$$y = \int 2x \, dx = x^2 + C,$$

where C is an arbitrary constant. In a sense, d/dx and $\int$ can be regarded as inverse operators, since if we set $d/dx = D$ and $\int = D^{-1}$, we immediately check that $DD^{-1} = 1$; on the other hand $D^{-1}D \neq 1$.

The analogous problem in finite differences is to be given Δy and to be required to find y. For example, if $\Delta y = 2x$, and the interval of differencing is h, we at once verify that

$$y = \frac{x^2}{h} - x + C.$$

Here the "constant of summation" C is not necessarily constant for all values of x; it need merely be constant for the values of x under consideration, that is, those separated by intervals of h. Hence we agree to the convention that *the constant of summation C shall represent an arbitrary function of period h*; it is clear that such a function C will have

$$C(a) = C(a + h) = C(a + 2h) = \dots .$$

We shall not, in this chapter, be primarily concerned with the problem of summation, that is, of finding y when given Δy. However, certain simple cases can be handled very easily. We agree to the convention

(9.1) $$x^{(r)} = x[x - h][x - 2h] \ldots [x - (r - 1)h].$$

The symbol $x^{(r)}$ is usually read "x to r factorial" and satisfies the recursion relation

(9.2) $$x^{(r)} = [x - (r - 1)h]x^{(r-1)}.$$

Putting $r = 1$, we obtain $x^{(1)} = x \cdot x^{(0)}$; thus, to keep (9.2) always valid, we see that we must define

(9.3) $$x^{(0)} = 1.$$

Also, putting $r = 0$ in (9.2) produces the result $x^{(0)} = (x + h)x^{(-1)}$. Again, to keep (9.2) valid, we define

(9.4) $$x^{(-1)} = \frac{1}{x + h},$$

and, continuing the same process, find

(9.5) $$x^{(-r)} = \frac{1}{(x + h)(x + 2h) \ldots (x + rh)}.$$

A natural extension of this notation leads us to define

(9.6) $(ax + b)^{(r)} = [ax + b][a(x - h) + b] \ldots [a(x - \overline{r - 1}\, h) + b].$

Elementary algebra now produces

LEMMA 9.1.
$$\Delta x^{(r)} = rhx^{(r-1)}.$$

COROLLARY 1. If $h = 1$,
$$\Delta x^{(r)} = rx^{(r-1)},$$
$$\Delta \binom{x}{r} = \frac{1}{r!}\Delta x^{(r)} = \binom{x}{r - 1}.$$

COROLLARY 2.
$$\Delta(ax + b)^{(r)} = arh(ax + b)^{(r-1)}.$$

If we introduce the operator Δ^{-1}, we can write the preceding results in the general form

(9.7) $$\Delta^{-1}(ax + b)^{(r-1)} = \frac{(ax + b)^{(r)}}{arh} + C.$$

Another direct and useful result is

LEMMA 9.2.
$$\Delta a^x = a^x(a^h - 1).$$

This may alternatively be stated as

$$\textbf{(9.8)} \qquad \Delta^{-1}a^x = \frac{a^x}{a^h - 1} + C.$$

We now connect the operator Δ^{-1} with the process of summation; suppose that $F(x)$ is a function having $\Delta F(x) = f(x)$. Then

$$F(a + h) - F(a) \qquad\qquad = \Delta F(a),$$
$$F(a + 2h) - F(a + h) \qquad\quad = \Delta F(a + h),$$
$$\cdots\cdots\cdots\cdots\cdots\cdots\cdots\cdots\cdots$$
$$F(a + nh) - F(a + \overline{n - 1}\,h) = \Delta F(a + \overline{n - 1}\,h),$$
$$F(a + \overline{n + 1}\,h) - F(a + nh) = \Delta F(a + nh).$$

Adding,

$$F(a + \overline{n + 1}\,h) - F(a) \qquad = \sum_{x=a}^{a+nh} f(x).$$

We usually write

$$\textbf{(9.9)} \qquad \sum_{x=a}^{a+nh} f(x) = [F(x)]_a^{a+(n+1)h},$$

where the square brackets indicate that $F(x)$ is to be evaluated between the upper and lower limits indicated. If we change the origin and scale so that a becomes the origin and the unit of differencing becomes 1, then we obtain

$$\textbf{(9.10)} \qquad \sum_0^n f(x) = [F(x)]_0^{n+1} = [\Delta^{-1}f(x)]_0^{n+1}.$$

This is completely analogous to the usual formula for definite integration (*but note the difference in the upper limits*).

EXAMPLE 9.1.   Evaluate $\displaystyle\sum_{x=0}^{n} (x^2 + 3x - 2)$.

$$\sum_0^n (x^2 + 3x - 2) = \sum_{x=0}^{n} [x(x - 1) + 4x - 2]$$
$$= \sum_{x=0}^{n} [x^{(2)} + 4x^{(1)} - 2x^{(0)}]$$
$$= \left[\frac{x^{(3)}}{3} + \frac{4x^{(2)}}{2} - \frac{2x^{(1)}}{1}\right]_0^{n+1}$$
$$= \tfrac{1}{3}(n + 1)^{(3)} + 2(n + 1)^{(2)} - 2(n + 1)^{(1)}$$
$$= \tfrac{1}{3}(n + 1)(n^2 + 5n - 6)$$
$$= \tfrac{1}{3}(n^2 - 1)(n + 6).$$

EXAMPLE 9.2. Find the sum of n terms of the series

$$\frac{1}{1 \cdot 4} + \frac{1}{4 \cdot 7} + \frac{1}{7 \cdot 10} + \dots$$

The sum required is

$$\sum_{x=0}^{n-1} \frac{1}{(3x+1)(3x+4)} = \sum_{x=0}^{n-1} (3x-2)^{(-2)}$$

$$= \left[\frac{(3x-2)^{(-1)}}{-3} \right]_0^n$$

$$= \tfrac{1}{3} \left[\frac{-1}{3n+1} + 1 \right]$$

$$= \tfrac{1}{3} \left[1 - \frac{1}{3n+1} \right] = \frac{n}{3n+1}.$$

More complicated series than those in Examples 9.2 and 9.3 can be summed by developing a formula for "summation by parts". Indeed, let $h = 1$ and

$$F(x) = \Delta^{-1} f(x), \qquad G(x) = \Delta^{-1} g(x).$$

Then

$$\Delta[F(x)G(x)] = F(x+1)G(x+1) - F(x)G(x)$$
$$= F(x+1)\, \Delta G(x) + G(x)\, \Delta F(x).$$

Hence

$$\sum_0^n G(x)\, \Delta F(x) = \sum_0^n \{ \Delta[F(x)G(x)] - F(x+1)\, \Delta G(x) \}.$$

This gives us

(9.11) $\qquad \displaystyle\sum_0^n G(x)\, \Delta F(x) = [F(x)G(x)]_0^{n+1} - \sum_0^n F(x+1)\Delta G(x).$

Formula (9.11) is the finite-difference form of the formula for integration by parts.

EXAMPLE 9.3. Find the sum of n terms of the double arithmetic-geometric progression

$$1 \cdot 2 \cdot 3 + 2 \cdot 3 \cdot 3^2 + 3 \cdot 4 \cdot 3^3 + 4 \cdot 5 \cdot 3^4 + 5 \cdot 6 \cdot 3^5 + \dots$$

$$\sum_0^{n-1} (x+1)(x+2)3^{x+1} = \sum_0^{n-1} (x+2)^{(2)}3^{x+1}$$

$$= \sum_0^{n-1} (x+2)^{(2)}\, \Delta \frac{3^{x+1}}{2}$$

$$= \left[(x+2)^{(2)}\, \frac{3^{x+1}}{2} \right]_0^n - \sum_0^{n-1} 3^{x+2}(x+2)^{(1)}.$$

The required sum then is

$$(n + 2)^{(2)} \frac{3^{n+1}}{2} - 3 - \sum_{0}^{n-1} (x + 2)^{(1)} \Delta \frac{3^{x+2}}{2}$$

$$= (n + 2)^{(2)} \frac{3^{n+1}}{2} - 3 - \left[(x + 2)^{(1)} \frac{3^{x+2}}{2} \right]_{0}^{n} + \sum_{0}^{n-1} \frac{3^{x+3}}{2}$$

$$= (n + 2)^{(2)} \frac{3^{n+1}}{2} - 3 - (n + 2)^{(1)} \frac{3^{n+2}}{2} + 3^2 + \frac{3^{n+3}}{4} - \frac{27}{4}$$

$$= \frac{3^{n+1}}{4} (2n^2 + 1) - \tfrac{3}{4} = \tfrac{3}{4}[3^n(2n^2 + 1) - 1].$$

We conclude this section with an instructive example which simplifies summations like that of Example 9.3.

EXAMPLE 9.4. Prove

$$\Delta^{-1} a^x f(x) = \frac{a^x}{a - 1} \left[1 - \frac{a}{a - 1} \Delta + \left(\frac{a}{a - 1} \right)^2 \Delta^2 - \ldots \right] f(x).$$

We have

$$\Delta^{-1} a^x f(x) = (E - 1)^{-1} a^x f(x).$$

Let us now introduce two partial enlargement operators E_1 and E_2 which affect *only* $f(x)$ and a^x respectively. Thus

$$E_1 f(x) = f(x + 1), \quad \text{but} \quad E_1 a^x = a^x;$$
$$E_2 a^x = a^{x+1}, \quad \text{but} \quad E_2 f(x) = f(x).$$

Let Δ_1 and Δ_2 be the corresponding difference operators. Then $E = E_1 E_2$ and

$$\Delta^{-1} a^x f(x) = (E_1 E_2 - 1)^{-1} a^x f(x)$$

$$= (\Delta_1 \Delta_2 + \Delta_1 + \Delta_2)^{-1} a^x f(x)$$

$$= \Delta_2^{-1} \left[1 + \frac{\Delta_1 E_2}{\Delta_2} \right]^{-1} a^x f(x)$$

$$= \left[\Delta_2^{-1} - \frac{\Delta_1}{\Delta_2^2} E_2 + \frac{\Delta_1^2}{\Delta_2^3} E_2^2 - \ldots \right] a^x f(x)$$

$$= \left[\frac{a^x}{a - 1} - \frac{a^{x+1}}{(a - 1)^2} \Delta_1 + \frac{a^{x+2}}{(a - 1)^3} \Delta_1^2 - \ldots \right] f(x).$$

Since Δ_1 now operates only on $f(x)$, we can dispense with the subscript and conclude with the result

$$\Delta^{-1} a^x f(x) = \frac{a^x}{a - 1} \left[1 - \frac{a}{a - 1} \Delta + \left(\frac{a}{a - 1} \right)^2 \Delta^2 - \ldots \right] f(x).$$

EXERCISES

1. Evaluate

$$\sum_1^n x, \quad \sum_1^n x^2, \quad \sum_1^n x^3.$$

2. By breaking $\dfrac{x^2}{(x+2)(x+3)}$ into partial fractions, evaluate

$$\sum_1^n \frac{x^2}{(x+2)(x+3)} \, 4^x.$$

3. Prove that

$$\Delta^n x^{-1} = (-)^n \frac{n!}{x(x+1)\dots(x+n)} \, .$$

Deduce the sum of the series

$$\frac{\binom{n}{0}}{x} - \frac{\binom{n}{1}}{x+1} + \frac{\binom{n}{2}}{x+2} - \frac{\binom{n}{3}}{x+3} + \dots$$

4. Sum the following series:

  (a) $\displaystyle\sum_0^n (2x+1)(2x+3),$ (b) $\displaystyle\sum_0^n \frac{1}{(2x+1)(2x+3)},$

  (c) $\displaystyle\sum_0^\infty \frac{1}{(2x+1)(2x+3)},$ (d) $\displaystyle\sum_0^n (2x+1)(2x+3)5^x,$

  (e) $\displaystyle\sum_0^n (1+2x+3x^2+4x^3),$ (f) $\displaystyle\sum_0^n (1+x)7^x.$

5. Find $\Delta \dfrac{2^x(x+1)!}{(2x+3)!} \, .$

6. Find $\Delta^{-1}\left[\dfrac{x^3}{(x+1)(x+2)(x+3)}\right].$

7. Use Example 9.4 to evaluate

$$\sum_1^n 5^x(3 - 2x + 4x^2 - x^3).$$

2. DIFFERENTIAL AND DIFFERENCE EQUATIONS

In this section we continue to stress the analogy between the processes of differentiation and differencing. Let us consider the origin of differential and difference equations.

Suppose a one-parameter family of curves $y = cx^2$ is given; then it is easy to find $dy/dx = 2cx$. Eliminating the parameter leaves us with a differential equation

$$\frac{dy}{dx} - 2\frac{y}{x} = 0.$$

In an exactly similar manner, suppose $y = cx^2$ is given (c a "constant", that is, a function of period $h = 1$), and suppose we form $\Delta y = c(2x + 1)$. Again, eliminating the parameter c, we obtain

$$\Delta y = \frac{y}{x^2}(2x + 1),$$

that is,

$$\Delta y = \frac{2y}{x} + \frac{y}{x^2}.$$

Just as $dy/dx = 2y/x$ is the *differential equation* corresponding to the one-parameter system $y = cx^2$, so

$$\Delta y = \frac{2y}{x} + \frac{y}{x^2}$$

is the *difference equation* corresponding to this same one-parameter family.

EXAMPLE 9.5. Form the differential and difference equations corresponding to the two-parameter family

$$y = ax^2 - bx.$$

We at once deduce

$$y' = 2ax - b,$$
$$y'' = 2a.$$

Eliminating a and b, we obtain

$$\begin{vmatrix} y & x^2 & x \\ y' & 2x & 1 \\ y'' & 2 & 0 \end{vmatrix} = 0,$$

that is, $x^2 y'' - 2xy' + 2y = 0$.

A similar procedure yields

$$\Delta y = a(2x + 1) - b,$$
$$\Delta^2 y = 2a.$$

Hence

$$\begin{vmatrix} y & x^2 & x \\ \Delta y & 2x + 1 & 1 \\ \Delta^2 y & 2 & 0 \end{vmatrix} = 0,$$

that is,

$$x^2 \Delta^2 y + x\,\Delta^2 y - 2x\,\Delta y + 2y = 0.$$

Thus the system of curves $y = ax^2 - bx$ is equivalent to the differential equation

$$x^2 D^2 y - 2x Dy + 2y = 0,$$

or to the difference equation

$$(x^2 + x)\Delta^2 y - 2x\,\Delta y + 2y = 0.$$

The difference equation might equally well be given in terms of E as

$$(x^2 + x)y_{x+2} - (2x^2 + 4x)y_{x+1} + (x^2 + 3x + 2)y_x = 0.$$

EXERCISES

1. Form the differential and difference equations corresponding to

   (a) $y = c^2 x^2 + c$,　　　　　　　　(b) $y = c2^x + c^2$,

   (c) $y = c + c^2 x + c^3 x^2$.

2. Form the differential and difference equations corresponding to

   (a) $y = \dfrac{a}{x} + b$,　　　　　　　　(b) $y = c2^x + d3^x$,

   (c) $y = c2^x + d3^x + b$,　　　　　　(d) $y = a + bx + cx^3$.

3. Form the difference equation (in terms of f_{x+1} and f_x) which is generated by $f_x = cx + c^4$.

4. Form the difference equation (in terms of f_x, f_{x+1}, f_{x+2}) which is generated by $f_x = c_1 2^x + c_2 x$.

3. EQUATIONS WITH CONSTANT COEFFICIENTS

The simplest type of differential equation, the linear homogeneous equation with constant coefficients, has the general form

(9.12)　　　　　　$(D^n + a_{n-1}D^{n-1} + \dots + a_1 D + a_0)y = 0,$

where the a_i are constants. The solution is easily obtained; indeed, let $y = e^{mx}$ be a solution. Then, by substitution,

$$e^{mx}(m^n + a_{n-1}m^{n-1} + \dots + a_1 m + a_0) = 0,$$

that is, m must satisfy the auxiliary equation

(9.13)　　　　　　$m^n + a_{n-1}m^{n-1} + \dots + a_1 m + a_0 = 0.$

If the roots $m_1, m_2, \dots, m_n$ of (9.13) are all distinct, then the most general solution of (9.12) is

(9.14)　　　　　　　　　　$\displaystyle\sum_{j=1}^{n} c_j e^{m_j x},$

where the c_j are arbitrary constants; if the m_j are not all distinct, the solution is slightly more complicated, and the result need not be quoted here.

Let us now treat the linear homogeneous difference equation

$$(9.15) \qquad (\Delta^n + a_{n-1}\Delta^{n-1} + \dots + a_1\Delta + a_0)f_x = 0$$

in a similar fashion. By replacing Δ by $E - 1$, Equation (9.15) can be written in the equivalent form

$$(9.16) \qquad (E^n + b_{n-1}E^{n-1} + \dots + b_1E + b_0)f_x = 0.$$

Let m^x be a solution of (9.16); substituting,

$$m^{x+n} + b_{n-1}m^{x+n-1} + \dots + b_1m^{x+1} + b_0m^x = 0.$$

Hence m must satisfy the auxiliary equation

$$(9.17) \qquad m^n + b_{n-1}m^{n-1} + \dots + b_1m + b_0 = 0.$$

Again, let us suppose that $m_1, m_2, \dots, m_n$ are all distinct; then

$$(9.18) \qquad \sum_{j=1}^{n} c_j m_j^x,$$

where the c_j are arbitrary constants, is the most general solution of (9.16). If the m_j are not all distinct, a modification, similar to that needed for the corresponding differential equation problem, is required.

EXAMPLE 9.6. Solve

$$(D^2 - 2D - 8)y = 0 \quad \text{and} \quad (E^2 - 2E - 8)y = 0.$$

For both equations, the auxiliary equation is $m^2 - 2m - 8 = 0$ with roots -2 and 4. Hence the solution of the differential equation is $c_1e^{-2x} + c_2e^{4x}$ and that of the difference equation is $c_1(-2)^x + c_2 4^x$.

It is worth noting that the difference equation might have been given in either of the two equivalent forms

$$y_{x+2} - 2y_{x+1} - 8y_x = 0,$$
$$(\Delta^2 - 9)y = 0.$$

EXERCISES

1. Prove that if $g(x)$ and $h(x)$ are two functions satisfying Equation (9.12), then $A\,g(x) + B\,h(x)$ also satisfies (9.12).

2. Repeat Exercise 1 for Equation (9.15).

3. Show that the general solution of

$$(D^3 + 3D^2 - 9D + 5)y = 0$$

is given by $(c_1 + c_2 x)e^x + c_3 e^{-5x}$. Show also that the general solution of

$$(E^3 + 3E^2 - 9E + 5)y = 0$$

is given by $(c_1 + c_2 x)1^x + c_3(-5)^x$.

4. Obtain the general solutions of the following equations:

    (a) $(D^3 - 6D^2 + 11D - 6)y = 0$,

    (b) $(E^3 - 6E^2 + 11E - 6)y = 0$,

    (c) $(E^3 + 3E^2 - 4E - 12)y = 0$,

    (d) $(E^4 - 13E^2 + 36)y = 0$.

5. Solve the equation

$$(E^3 + E^2 - 9E - 9)y = 5(2^x)$$

by solving the homogeneous equation

$$(E^3 + E^2 - 9E - 9)y = 0,$$

and noting that $a \cdot 2^x$ is a *particular integral* of the given equation for a suitably chosen.

6. Solve the equations:

    (a) $(\Delta^2 - 3\Delta + 2)y = 0$,

    (b) $(\Delta^3 + 3\Delta^2 - 13\Delta - 15)y = 0$,

    (c) $f(x + 3) - f(x + 2) - 14f(x + 1) + 24f(x) = 0$,

    (d) $(\Delta - 4)(\Delta - 5)(\Delta + 2)(\Delta + 1)f(x) = 0$,

    (e) $(\Delta - 2)^2(\Delta - 5)f(x) = 0$,

    (f) $f(x + 2) + f(x + 1) = 56f(x)$.

4. NUMERICAL SOLUTION OF DIFFERENCE EQUATIONS

Let us write down the most general nth order difference equation

(9.19) $(A_n E^n + A_{n-1}E^{n-1} + \ldots + A_1 E + A_0)f_x = A,$

where $A, A_0, A_1, \ldots, A_n$ are functions of x. The homogeneous equation with constant coefficients (9.16) is the special case of (9.19) when $A_n = 1$, $A = 0$, and $A_0, A_1, \ldots, A_{n-1}$ are independent of x. Since the solution will, in general, involve n arbitrary constants, we shall be able to impose n conditions upon the solution.

Now if h is the interval of differencing, (9.19) merely relates successive functional values $f_x, f_{x+h}, f_{x+2h}, \ldots$. Hence (9.19) reduces to the solution of a number of simultaneous linear equations, provided we require not the general analytic solution, but merely a numerical tabulation of functional values at $x, x+h, x+2h, \ldots$.

EXAMPLE 9.7. Solve the equation

$$f(x+2h) + f(x+h) - 12f(x) = 10x$$

in the interval $(0, 1)$ with interval of differencing $h = .1$ and subject to the boundary values $f(0) = 0, f(1) = 50$.

Denoting the ordinates at $0, .1, .2, \ldots$ by names $f_0, f_1, f_2, \ldots$, we find that we must deal with the system of equations

$$
\begin{aligned}
f_1 + f_2 \qquad\quad &= 0, \\
-12f_1 + f_2 + f_3 &= 1, \\
-12f_2 + f_3 + f_4 &= 2, \\
-12f_3 + f_4 + f_5 &= 3, \\
\cdots\cdots&\cdots\cdots \\
-12f_7 + f_8 + f_9 &= 7, \\
-12f_8 + f_9 + 50 &= 8.
\end{aligned}
$$

This is a set of nine equations in nine unknowns and, as such, can be solved explicitly to give values of f_x for $x = 0(.1)1$.

In this particular case, we obtain

$$
\begin{aligned}
f_2 &= -f_1, \\
f_3 &= 1 + 13f_1, \\
f_4 &= 1 - 25f_1, \\
f_5 &= 14 + 181f_1, \\
f_6 &= 2 - 481f_1, \\
f_7 &= 171 + 2653f_1, \\
f_8 &= -141 - 8425f_1, \\
f_9 &= 2200 + 40261f_1.
\end{aligned}
$$

Then, from $f_9 - 12f_8 = -42$, we obtain

$$141361f_1 = 3934,$$

that is,

$$f_1 = -.02782945791.$$

The other values can now all be computed (note, however, that this method has certain disadvantages; since some of the coefficients multiplying are f_1 as large as 40000, the retention of eleven decimals in f_1 is necessary in order to have a tabulated solution good to six decimal places). The complete solution is given in Table 9.1 (entries have been rounded to five decimals after being computed from the eleven-decimal value of f_1).

Table 9.1. SOLUTION OF

$$f(x + 2h) + f(x + h) - 12f(x) = 10x$$

x	$f(x)$
0	0.00000
.1	−.02783
.2	+.02783
.3	+.63822
.4	+1.69574
.5	+8.96287
.6	+15.38597
.7	+97.16845
.8	+93.46318
.9	+1079.55820
1.0	50.00000

EXERCISES

1. Show that if we change the scale so that $h = 1$, the equation in Example 9.7 takes the form
$$f(x + 2) + f(x + 1) - 12f(x) = x$$
in the interval (0, 10).

2. Show that $ax + b$ is a particular integral of Exercise 1 for a and b suitably chosen (substitute in the difference equation); thence deduce that the explicit solution of Exercise 1 is
$$-.00173864886(-4)^x + .03173864886(3)^x - .1x - .03.$$

3. Use Exercise 2 to verify that
$$f_1 = -.0278294579.$$

Verify also the entry for f_9 in Table 9.1

4. Tabulate the solution of

$$f(x + 2h) - 5f(x + h) + 6f(x) = 20x$$

in 0(.1)1 if $h = .1$, and the initial conditions are $f(0) = 0, f(1) = 100$.

5. Tabulate the solution of

$$f(x + 2h) - 6f(x + h) + 8f(x) = 15x$$

in 0(.2)1 if $h = .2$ and the initial conditions are $f(0) = 0, f(1) = 100$.

5. SOLUTION OF DIFFERENCE EQUATIONS BY RELAXATION

Usually the general solution of the difference equation (9.19) will either not be available or, if theoretically attainable, will not be in a form suitable for numerical computation. Consequently, we shall consider the solution of the difference equation as numerically equivalent to solving a system of simultaneous linear equations, as explained and illustrated in the last section. However, the systems of linear equations arising in solving difference equations are frequently of a very special type (cf. Example 9.7) where a great deal of symmetry exists in the coefficient pattern. Because of this fact, many difference equations (that is, systems of simultaneous linear equations) can be quickly handled numerically by a relaxation method.

EXAMPLE 9.8. Consider Example 9.7, in which the typical equation might be written

$$f_{i+1} + f_i - 12f_{i-1} = i - 1 \qquad (i = 1, \dots, 9)$$

with boundary values $f_0 = 0, f_{10} = 50$. In reality, there is only one general relaxation operator connected with this set of equations; it is usually designated diagrammatically as

and called a "point relaxation" operator. The symbolism indicates that a change of 1 unit in f_i produces changes of 1, 1, and -12 units respectively in R_{i-1}, R_i, and R_{i+1}; *these changes are independent of i.* Note that f_0 and f_{10} are not permitted to change. Note also that the point indicated by the solid circle shows the value of f_i which changes.

Group operators can be indicated on a schema just as readily; for example, combining the point operator just given with itself (for other points) would give the operator

Table 9.2. OPERATIONS TABLE FOR EXAMPLE 9.9

f_0	f_1	f_2	f_3	f_4	f_5	f_6	f_7	f_8	f_9	f_{10}
0										500
									150	
								50		
							14			
									13	
						3				
								1		
							1			
								1		
									1	
0	0	0	0	0	0	30	150	520	1640	500
						6				
							6			
								6		
					5					
									3	
						4				
							3			
								1		
			−3							
				−4						
		−2								
			−2							
		−1								
	−1									
				−1						
					−1					
0	−1	−3	−5	−5	4	40	159	527	1643	500

This operator is the result of unit changes in 6 of the f-values, and is called a line operator. Naturally, line operators may involve any number of f-values.

Example 9.7 has been solved without relaxation; actually, problems such as Example 9.7 which involve negative roots of the auxiliary equation will have solutions which oscillate wildly and are not very suitable for relaxation methods. Relaxation works better when the solution changes more continuously as in

EXAMPLE 9.9. Solve the equation

$$2f(x + 2) - 7f(x + 1) + 3f(x) = x$$

given that $f_0 = 0, f_{10} = 500$.

Table 9.3. RESIDUAL TABLE FOR EXAMPLE 9.9

R_1	R_2	R_3	R_4	R_5	R_6	R_7	R_8	R_9
0	−1	−2	−3	−4	−5	−6	−7	992
							293	−58
						94	−57	92
					23	−4	−15	
							11	1
				2	2	5		
						7	4	4
					4	0	7	
						2	0	7
							2	0
0	−10	−20	−30	20	40	20	20	0
				32	−2	38		
					10	−4	38	
						8	−4	18
							2	−3
			−20	−3	25			
				5	−3	20		
					3	−1	11	
						1	4	0
	−16	1	−29					
		−7	−1	−7				
−4	−2	−13						
	−6	1	−7					
−6	1	−2						
1	−2							
		−4	0	−10				
			−2	−3	0			
1	−2	−4	−2	−3	0	1	4	0

The system of linear equations is

$$-7f_1 + 2f_2 = 0,$$
$$3f_1 - 7f_2 + 2f_3 = 1,$$
$$3f_2 - 7f_3 + 2f_4 = 2,$$
$$\cdots\cdots\cdots\cdots\cdots\cdots\cdots\cdots$$
$$3f_7 - 7f_8 + 2f_9 = 7,$$
$$3f_8 - 7f_9 + 1000 = 8.$$

The relaxation operator common to this system of equations is

$$\begin{array}{ccc} 2 & -7 & 3 \\ \circ & \bullet & \circ \end{array}$$
$$f_i$$

Let us apply this point operator (Tables 9.2 and 9.3).

A number of points become clear from a careful study of Tables 9.2 and 9.3. First, we see that "over-relaxation" is usually a good idea; it is no use to knock out one residual exactly if it is to be partly restored in the very next step; consequently, it is often desirable to perform an operation that alters the sign of the residual (this is done in the first line of Table 9.3, and, with a little experience, one would over-relax even further in that line). Secondly, it is clear that using line relaxation operators would speed up the work. For example, the first three lines of the table following the multiplication by ten might be simultaneously effected by using the line operator

$$\begin{array}{ccccc} 2 & -7 & 3 & & \\ & 2 & -7 & 3 & \\ & & 2 & -7 & 3 \\ 2 & -5 & -2 & -4 & 3 \end{array}$$

A certain type of line operator known as a "wedge operator" is also of considerable use; this is a type of operator in which almost the entire effect is concentrated in one residual (and can thereby liquidate one large boundary residual such as the entry 992 in Example 9.9). For example, the operator

$$\begin{array}{ccccc} 2 & -7 & 3 & & \\ & 6 & -21 & 9 & \\ & & 18 & -63 & 27 \\ 2 & -1 & 0 & -54 & 27 \end{array}$$

is an excellent wedge operator for Example 9.9. If we apply a multiple of 18 times this operator to the initial residuals of Table 9.3, we obtain, *in one step*, the drastically reduced set of residuals

$$0, \quad -1, \quad -2, \quad -3, \quad -4, \quad 31, \quad -24, \quad -7, \quad 20.$$

This already produces a respectable approximate solution

$$(0, \quad 0, \quad 0, \quad 0, \quad 0, \quad 0, \quad 18, \quad 54, \quad 162).$$

Minor adjustments with point and line operators can then be used to further whittle down the residuals.

EXERCISES

1. Continue Example 9.9 to another decimal place, giving

i	1	2	3	4	5	6	7	8	9
$10^2 f_i$	−10	−36	−61	−60	31	397	1593	5278	16434
$10^2 R_i$	−2	0	−1	−1	−3	0	−4	1	−4

2. Show that the exact solution of Example 9.9 is given by

$$A\left(\frac{1}{2}\right)^x + B3^x - \frac{x}{2} + \frac{3}{4}.$$

Determine A and B, and use the theoretical solution to check the values obtained in Exercise 1.

3. Use relaxation procedures to obtain numerical solutions for the following equations, given the boundary values.

(a) $3f_{x+2} - 10f_{x+1} + 3f_x = 2x$ $(f_0 = 10, f_{10} = 1000)$;

(b) $2f_{x+2} - 7f_{x+1} + 2f_x = 2^x$ $(f_0 = 2,\ f_{10} = 100)$;

(c) $f_{x+3} - 6f_{x+2} + 11f_{x+1} - 5f_x = 0$ $(f_0 = 10, f_{10} = 1000)$;

(d) $f_{x+3} - 5f_{x+2} + 10f_{x+1} - 4f_x = 3^x$ $(f_0 = 50, f_{10} = 5000)$.

4. (Cf. Exercise 1 of Section 4.) For the following equations, change the scale so as to obtain new equations with $h = 1$; thence solve the equations.

(a) $4f_{x+2h} - 9f_{x+h} + 2f_x = 50x$ $(f_0 = 0, f_1 = 1000, h = .1)$;

(b) $3f_{x+2h} - 10f_{x+h} + 2f_x = 10x$ $(f_0 = 0, f_1 = 500, h = .2)$.

CHAPTER TEN

Solution of Differential Equations by Difference Equation Methods

1. RELATIONS BETWEEN DERIVATIVES AND DIFFERENCES

Suppose that the interval of differencing is h, that is,

$$f_0 = f(x_0), \quad f_1 = f(x_0 + h), \quad f_2 = f(x_0 + 2h), \quad \ldots;$$

then we may use Taylor's series to write

$$f_1 = f_0 + \frac{h}{1!} f_0' + \frac{h^2}{2!} f_0'' + \frac{h^3}{3!} f_0''' + \cdots$$

and

$$f_{-1} = f_0 - \frac{h}{1!} f_0' + \frac{h^2}{2!} f_0'' - \frac{h^3}{3!} f_0''' + \cdots$$

Subtracting, we obtain

(10.1)
$$f_1 - f_{-1} = 2\left[\frac{h}{1!} f_0' + \frac{h^3}{3!} f_0''' + \cdots\right].$$

The left-hand side of (10.1) can be written in terms of f_0 as

$$(E - E^{-1})f_0 = 2\mu\delta f_0.$$

The right-hand side of (10.1) can be written as

$$2\left[\frac{hD}{1!} + \frac{h^3 D^3}{3!} + \cdots\right]f_0 = 2(\sinh hD)f_0.$$

Equating these two expressions yields the important operational formula

(10.2) $\mu\delta = \sinh hD.$

This formula can also be written in the form $hD = \sinh^{-1}\mu\delta.$

If we assume that hD is sufficiently small that powers higher than the first may be neglected, then (10.2) assumes the simple form

(10.3) $\mu\delta = hD.$

In (10.3), we have a formula connecting first differences and first derivatives (neglecting higher derivatives or differences). If we raise (10.3) to successive powers, always neglecting higher differences and derivatives, we obtain

$$h^2 D^2 = \mu^2\delta^2 = \left(1 + \frac{\delta^2}{4}\right)\delta^2 = \delta^2.$$

Thus we have

(10.4) $\delta^2 = h^2 D^2;$

and similarly,

(10.5) $\mu\delta^3 = h^3 D^3,$

(10.6) $\delta^4 = h^4 D^4.$

Formulae (10.3), (10.4), (10.5), (10.6) will suffice for our purposes; if we apply these four operator formulae to f_0, they take the forms

(10.7) $f_0' = \frac{1}{2h}(f_1 - f_{-1}),$

(10.8) $f_0'' = \frac{1}{h^2}(f_1 - 2f_0 + f_{-1}),$

(10.9) $f_0''' = \frac{1}{2h^3}(f_2 - 2f_1 + 2f_{-1} - f_{-2}),$

(10.10) $f_0^{iv} = \frac{1}{h^4}(f_2 - 4f_1 + 6f_0 - 4f_{-1} + f_{-2}).$

It is well to recall again that (10.7), etc., are approximations, in that higher differences and derivatives have been neglected.

If, for any reason, we want to put Formulae (10.3), etc., in a form involving only subsequent ordinates, this can readily be done.

EXAMPLE 10.1. Express f_0' in terms of f_0 and f_1.

Here, since two ordinates are given, we assume $\Delta^2 = 0$; then

$$hD = \mu\delta = \frac{1}{2}(E - E^{-1}) = \frac{1}{2}\left(1 + \Delta - \frac{1}{1+\Delta}\right)$$

$$= \frac{1}{2}(1 + \Delta - 1 + \Delta) = E - 1.$$

Hence

$$f_0' = \frac{1}{h}(f_1 - f_0).$$

EXAMPLE 10.2. Express f_0'' in terms of f_0, f_1, f_2, f_3.

Since four ordinates are given, assume $\Delta^4 = 0$; then

$$h^2 D^2 = \delta^2 = E - 2 + E^{-1} = (1 + \Delta) - 2 + (1 - \Delta + \Delta^2 - \Delta^3)$$
$$= \Delta^2 - \Delta^3 = 2 - 5E + 4E^2 - E^3.$$

Hence

$$f_0'' = \frac{1}{h^2}(2f_0 - 5f_1 + 4f_2 - f_3).$$

EXERCISES

1. Prove that $e^{hD} = E$, and use this result to give (10.2).

2. Deduce Formulae (10.7) and (10.8) from the result

$$hD = \log E = \log(1 + \Delta).$$

3. Show that (10.3) is equivalent to the assumption that the tangent at (x_0, f_0) is parallel to the chord joining (x_1, f_1) and (x_{-1}, f_{-1}). Illustrate on a diagram.

4. Prove (10.8), (10.9), (10.10), by using the Taylor expansions for f_1, f_2, f_{-1}, f_{-2} (*this method shows exactly what derivatives are neglected*).

5. Prove the following formulae:

(a) $f_0'' = \frac{1}{h^3}(f_3 - 3f_2 + 3f_1 - f_0),$

(b) $f_0'' = \frac{1}{2h^3}(-5f_0 + 18f_1 - 24f_2 + 14f_3 - 3f_4).$

Explain the different hypotheses underlying these two formulae.

6. Prove the result

$$f_0^v = \frac{1}{2h^5}(f_3 - 4f_2 + 5f_1 - 5f_{-1} + 4f_{-2} - f_{-3}).$$

7. Show that the differential equation

$$y'' + 5y' - 3y = 0,$$

when written for the specific point (x_0, y_0), yields the approximating difference equation

$$(2 + 5h)y_1 + (2 - 5h)y_{-1} - (4 + 6h^2)y_0 = 0.$$

2. TRANSFORMATION OF DIFFERENTIAL EQUATIONS IN TO DIFFERENCE EQUATIONS

Suppose that a differential equation

(10.11)
$$G(x, y, y', y'', \ldots, y^{(n)}) = 0$$

is given; at the particular point (x_0, y_0), Equation (10.11) becomes

(10.12)
$$G(x_0, y_0, y_0', y_0'', \ldots, y_0^{(n)}) = 0.$$

Substituting from Formulae (10.7), (10.8), etc., we obtain a difference equation involving $y_0, y_1, y_{-1}, \ldots$. It can be handled by the methods of Chapter IX.

EXAMPLE 10.3.
$$x^2 y'' + (x - 2)y' - 3y = 10x.$$

Using (10.7) and (10.8) at the point (x_0, y_0) yields

$$\frac{x_0^2}{h^2}(y_1 - 2y_0 + y_{-1}) + \frac{(x_0 - 2)}{2h}(y_1 - y_{-1}) - 3y_0 = 10x_0,$$

that is,

$$y_1(2x_0^2 + hx_0 - 2h) + y_{-1}(2x_0^2 - hx_0 + 2h) - y_0(4x_0^2 + 6h^2) - 20h^2x_0 = 0.$$

If we wish a solution of this equation in the interval $(0, 1)$ with $h = .1$, $y(0) = 0$, $y(1) = 100$, we obtain a set of simultaneous linear equations by setting $x_0 = .1, .2, .3, \ldots, .9$, in turn. Thus

$$.21y_0 - .10y_1 - .17y_2 = .02,$$
$$.26y_1 - .22y_2 - .10y_3 = .04,$$
$$.35y_2 - .42y_3 + .01y_4 = .06,$$
$$.48y_3 - .70y_4 + .16y_5 = .08,$$
$$\cdots \cdots \cdots \cdots \cdots \cdots$$
$$1.73y_8 - 3.30y_9 + 1.51y_{10} = .18.$$

These nine equations determine the nine unknowns $y_1, \ldots, y_9$, and thus allow us to tabulate the solution in $0(.1)1$.

EXAMPLE 10.4. Solve the equation

$$y'' - 3y' - 10y = 10x,$$

in $(0, 1)$, subject to the conditions $y(0) = 0$, $y(1) = 100$.

We select $h = .1$ as a suitable mesh width, and obtain the approximating difference equation for the typical point (x_0, y_0) as

$$y_{-1}(2 + 3h) - y_0(4 + 20h^2) + y_1(2 - 3h) = 20h^2x_0.$$

Table 10.1. RELAXATION TABLE FOR EXAMPLE 10.4

0	0	0	0	0	0	0	0	0
			6	12	18	24	36	60
			−1	−2	−3			
				−2	−2	−2		
		2						
	1							
0	**10**	**20**	**50**	**80**	**130**	**220**	**360**	**600**
4		5					4	4
	2							
			−4					
1	1			−3				
						−2		
			−1				−1	−1
5	**13**	**25**	**45**	**77**	**130**	**218**	**363**	**603**

Table 10.2. RESIDUAL TABLE FOR EXAMPLE 10.4

−2	−4	−6	−8	−10	−12	−14	−16	16982
		1014	−488	−610	−732	166	584	62
		844	−408	−510	68	−524	124	
		−748	−10	−10	108	−144		
		4	−288					
	336							
168	−84	234						
168	**−84**	**234**	**−288**	**−10**	**108**	**−144**	**124**	**62**
0	8					−76	24	−14
	93	24	−173					
34	9	70						
		2	−5	−102				
9	−10	25						
			−56	24	39			
					5	8	−22	
		8	−14	1				
						−9	3	5
90	**−100**	**80**	**−140**	**10**	**50**	**−90**	**30**	**50**

This illustrates the fact that the positive and negative residuals should roughly balance out as we approach the solution; clearly, if there were a great preponderance of residuals of one sign, it would be an indication that our relaxation had leaned too far in one direction.

EXERCISES

1. Solve Example 10.3 by relaxation.

2. Show that the exact solution of Example 10.4 is

$$.67940e^{5x} - .97940e^{-2x} - x + .3.$$

Thence obtain the exact solution in (0, 1) as

x	y
0.0	0.000
0.1	0.518
0.2	1.290
0.3	2.507
0.4	4.480
0.5	7.716
0.6	13.051
0.7	21.857
0.8	36.396
0.9	60.396
1.0	100.000

3. In Example 10.4, construct the wedge operators

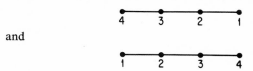

and

4. Verify that the values .51, 1.27, 2.47, 4.43, 7.64, 12.96, 21.75, 36.28, 60.30 give residuals of -3, -2, 18, -17, 33, -5, -6, -1, 4 (all times 10^{-3}) in Example 10.4.

5. Solve the following equations by relaxation in (0, 1) with $h = .1$; take the initial and final values of y to be 0 and 100 in each case.

   (a) $y'' - y' + 6y = 0$,

   (b) $y'' - 2y' + 3y = 10x$,

   (c) $y'' - 4y' + 4y = 10x$,

   (d) $y'' - 5y + 2 = e^x$.

6. Solve the equations

   (a) $xy'' - (x + 1)y' + 5y = 0$,
   (b) $(x + 2)y'' + x^2y' - 3y = 0$,

   in $(0, 1)$, subject to the conditions $y(0) = 10$, $y(1) = 500$. Take $h = .1$ in both cases.

7. Solve the equations

   (a) $(x^2 + 2)y' - 10xy = 10x$,
   (b) $3y' + 5xy = 10x$,

   in $(0, 1)$, given that $y(0) = 0$, $y(1) = 200$.

3. PARTIAL DERIVATIVES

In Section 1, we discussed the connection between derivatives and differences of a function $f(x)$ of a single variable x. We now consider a function $f(x, y)$ of two independent variables x and y (an analogous treatment will cover the case of more than two independent variables).

We shall need to distinguish between the operators μ, δ, E, ... , according as they operate on x or on y. We do this by means of the appropriate subscript. Thus (assuming that the interval of differencing is the same for both x and y), we have

$$E_x f_0 = E_x f(x_0, y_0) \quad = f(x_0 + h, y_0),$$
$$E_y f_0 = E_y f(x_0, y_0) \quad = f(x_0, y_0 + h),$$
$$E_x E_y f_0 = E_x E_y f(x_0, y_0) = f(x_0 + h, y_0 + h).$$

Let us now start at a typical point (x_0, y_0) in the x-y plane and consider the partial derivatives of $f(x, y)$ at this point. The neighbouring points can be represented on a mesh as in Figure 10.1. (The f-axis is, of course, perpendicular to the page, that is, the mesh in Figure 10.1 represents the domain over which f is defined.)

15	14	13	12	11
.	.	.	.	.
16	4	3	2	10
.	.	.	.	.
17	5	0	1	9
.	.	.	.	.
18	6	7	8	24
.	.	.	.	.
19	20	21	22	23
.	.	.	.	.

Fig. 10.1.

Any system of naming the points in Figure 10.1 is satisfactory; in our diagram, 0 designates (x_0, y_0), 1 designates $(x_0 + h, y_0)$, 14 designates $(x_0 - h, y_0 + 2h)$, etc. We at once apply Formula (10.7) to give

(10.13)
$$\frac{\partial f_0}{\partial x} = \left(\frac{\partial f}{\partial x}\right)_0 = \frac{1}{2h}(f_1 - f_5)$$

and

(10.14)
$$\frac{\partial f_0}{\partial y} = \left(\frac{\partial f}{\partial y}\right)_0 = \frac{1}{2h}(f_3 - f_7).$$

Similarly, (10.8) yields

(10.15)
$$\frac{\partial^2 f_0}{\partial x^2} = \frac{1}{h^2}(f_1 - 2f_0 + f_5)$$

and

(10.16)
$$\frac{\partial^2 f_0}{\partial y^2} = \frac{1}{h^2}(f_3 - 2f_0 + f_7).$$

The mixed partial derivative is a new result:

$$\frac{\partial^2 f_0}{\partial x\, \partial y} = D_x D_y f_0 = \frac{1}{h^2}(\mu_x \delta_x)(\mu_y \delta_y)f_0$$

$$= \frac{1}{4h^2}(E_x - E_x^{-1})(E_y - E_y^{-1})f_0$$

$$= \frac{1}{4h^2}(f_2 - f_4 - f_8 + f_6).$$

This is formula

(10.17)
$$\frac{\partial^2 f_0}{\partial x\, \partial y} = \frac{1}{4h^2}(f_2 - f_4 + f_6 - f_8).$$

The results for third and fourth derivatives are given in the Exercises; the pure derivatives follow from the results of Section 1, whereas the mixed derivatives can be developed using the two enlargement operators E_x and E_y, as in the derivation of (10.17).

Just as in the case of ordinary differential equations, we can replace partial differential equations by difference equations. We exemplify the procedure, which involves no new principle, in

EXAMPLE 10.5. Change

$$\frac{\partial^2 f}{\partial x^2} + \frac{\partial^2 f}{\partial y^2} = -500$$

into a difference equation.

Consider the equation at a specific point (x_0, y_0, f_0); then

$$\frac{\partial^2 f_0}{\partial x^2} + \frac{\partial^2 f_0}{\partial y^2} = -500.$$

Naming the points, in relation to 0, as in Figure 10.1, we have the result

$$\frac{1}{h^2}[f_1 - 2f_0 + f_5] + \frac{1}{h^2}[f_3 - 2f_0 + f_7] = -500.$$

$$f_1 + f_3 + f_5 + f_7 - 4f_0 = -500h^2.$$

An equation similar to this holds for every point in the region; we become independent of Figure 10.1 if we note that the left-hand side of this equation represents the sum of the functional values at the four points nearest 0, decreased by four times the functional value at the base point 0.

EXERCISES

1. In Figure 10.1, identify the points

$$f(x_0 - 2h, y_0 - h), \quad f(x_0 + h, y_0 - 2h),$$
$$E_x^2 E_y^2 f_0, \quad E_x^{-1} E_y^{-1} f_0, \quad E_x^{-1} f_0, \quad E_x E_y^{-2} f_0.$$

2. Obtain Formula (10.17) by using the Taylor expansions giving

$$f_2 = f(x + h, y + h), \quad f_4 = f(x - h, y + h),$$
$$f_6 = f(x - h, y - h), \quad f_8 = f(x + h, y - h).$$

3. Prove the third-derivative formulae:

$$\frac{\partial^3 f_0}{\partial x^3} = \frac{1}{2h^3}(f_9 - 2f_1 + 2f_5 - f_{17}),$$

$$\frac{\partial^3 f_0}{\partial y^3} = \frac{1}{2h^3}(f_{13} - 2f_3 + 2f_7 - f_{21}),$$

$$\frac{\partial^3 f_0}{\partial x \, \partial y^2} = \frac{1}{2h^3}[(f_2 - 2f_1 + f_8) - (f_4 - 2f_5 + f_6)],$$

$$\frac{\partial^3 f_0}{\partial x^2 \, \partial y} = \frac{1}{2h^3}[(f_2 - 2f_3 + f_4) - (f_8 - 2f_7 + f_6)].$$

4. Prove the fourth-derivative formulae:

$$\frac{\partial^4 f_0}{\partial x^4} = \frac{1}{h^4}(f_9 - 4f_1 + 6f_0 - 4f_5 + f_{17}),$$

$$\frac{\partial^4 f_0}{\partial y^4} = \frac{1}{h^4}(f_{13} - 4f_3 + 6f_0 - 4f_7 + f_{21}),$$

$$\frac{\partial^4 f_0}{\partial x \, \partial y^3} = \frac{1}{4h^4}[(f_{12} - 2f_2 + 2f_8 - f_{22}) - (f_{14} - 2f_4 + 2f_6 - f_{20})],$$

$$\frac{\partial^4 f_0}{\partial x^3 \, \partial y} = \frac{1}{4h^4}[(f_{10} - 2f_2 + 2f_4 - f_{16}) - (f_{24} - 2f_8 + 2f_6 - f_{18})],$$

$$\frac{\partial^4 f_0}{\partial x^2 \, \partial y^2} = \frac{1}{h^4}[(f_2 + f_4 + f_6 + f_8) - 2(f_1 + f_3 + f_5 + f_7) + 4f_0].$$

5. Change the following partial differential equations into difference equations at (x_0, y_0, f_0):

(a) $\dfrac{\partial f}{\partial x} + \dfrac{\partial f}{\partial y} = x + y,$

(b) $\dfrac{\partial^2 f}{\partial x^2} - 2\dfrac{\partial^2 f}{\partial x\, \partial y} + \dfrac{\partial^2 f}{\partial y^2} = -10,$

(c) $\dfrac{\partial^4 f}{\partial x^4} + 2\dfrac{\partial^4 f}{\partial x^2\, \partial y^2} + \dfrac{\partial^4 f}{\partial y^4} = 0,$

(d) $\dfrac{\partial^2 f}{\partial x^2} + \dfrac{A}{x}\dfrac{\partial f}{\partial y} + \dfrac{\partial^2 f}{\partial y^2} = 0,$

(e) $\dfrac{\partial^3 f}{\partial x^3} + \dfrac{\partial^3 f}{\partial y^3} = 0,$

(f) $\dfrac{\partial^2 f}{\partial x^2} + 2\dfrac{\partial^2 f}{\partial x\, \partial y} + 3\dfrac{\partial^2 f}{\partial y^2} = 0,$

(g) $\left[\dfrac{\partial^2 f}{\partial x^2} + \dfrac{\partial f}{\partial x}\right] - \left[\dfrac{\partial^2 f}{\partial y^2} + \dfrac{\partial f}{\partial y}\right] = 0.$

4. SOLUTION OF PARTIAL DIFFERENTIAL EQUATIONS BY DIFFERENCE METHODS

We exemplify the procedure, similar to that followed in Section 2, by solving

EXAMPLE 10.6.

$$\frac{\partial^2 f}{\partial x^2} + \frac{\partial^2 f}{\partial y^2} = -500.$$

It is hardly necessary to state that the Laplacian expression on the left-hand side of this equation is the commonest of all partial differential expressions.

We must be given some boundary values on f; suppose that we want to determine f within the square bounded by $x = 0$, $y = 0$, $x = 4$, $y = 4$. Suppose further that f is zero at every point on the boundary of this square.

Name the points of the square (we are taking $h = 1$) as shown in Figure 10.2; note that the system of nomenclature is completely arbitrary.

20	21	22	23	24
15	16	17	18	19
10	11	12	13	14
5	6	7	8	9
0	1	2	3	4

Fig. 10.2.

We are given the boundary values

$$f_0 = \ldots = f_4 = \ldots = f_{24} = \ldots = f_{20} = \ldots = f_5 = 0.$$

Using the result of Example 10.5, we see that there are nine equations for the nine unknowns

$$f_6, \ f_7, \ f_8, \ f_{11}, \ f_{12}, \ f_{13}, \ f_{16}, \ f_{17}, \ f_{18}.$$

These are

$$R_{16} = f_{15} + f_{21} + f_{17} + f_{11} - 4f_{16} + 500 = 0,$$
$$R_{17} = f_{16} + f_{18} + f_{22} + f_{12} - 4f_{17} + 500 = 0,$$
$$\cdot \ \cdot \ \cdot \ \cdot \ \cdot \ \cdot \ \cdot \ \cdot \ \cdot \ \cdot \ \cdot \ \cdot \ \cdot \ \cdot \ \cdot \ \cdot \ \cdot \ \cdot$$
$$R_8 \ = f_7 + f_3 + f_9 + f_{13} - 4f_8 + 500 \quad = 0.$$

We now write down the relaxation operator connected with any point f_i. It is shown in Figure 10.3. This symbolism denotes the fact that a change of

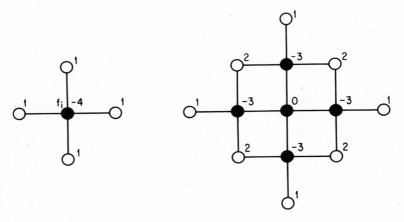

Fig. 10.3.   Fig. 10.4.

one unit in f_i causes a change of -4 units in R_i and one unit in each of the four "adjacent" residuals (adjacent, in the sense of nomenclature). We can combine point relaxation operators such as the one just given to obtain "block operators"; for example, a set of five relaxations gives a block. It is shown in Figure 10.4. Other possibilities are shown in Figures 10.5 and 10.6. It need hardly be mentioned that symmetrical blocks are the most desirable if there is any sort of regularity in the problem.

So far, we have kept the relaxation table and the residual table separate, largely because this procedure makes the steps in the solution clearer. The student, however, has probably noticed that one table, with double-entry columns, would have been physically handier. In problems such as the present one, ease of physical attack becomes paramount and we arrange our work under the following convention: the relaxation and residual tables are combined; *relaxations are written to the left of the point and residuals are written to the right.* Starting from the trial solution

$$f_{16} = f_{17} = f_{18} = f_{11} = f_{12} = f_{13} = f_6 = f_7 = f_8 = 0,$$

we construct the joint table shown in Figure 10.7. In the first step, we apply a point relaxation to eliminate R_{12}. Then we use 94 times the block of Figure

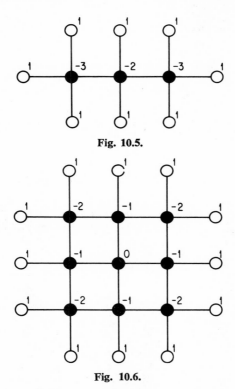

Fig. 10.5.

Fig. 10.6.

10.4; the coefficient 94 is selected so that the residuals at points 16, 17, etc., will be in the ratio 2 : 1 and thus be almost entirely liquidated by applying the block of Figure 10.6. The final solution is

$$f_{16} = f_6 = f_{18} = f_8 = 344,$$
$$f_{17} = f_{11} = f_7 = f_{13} = 438, \qquad f_{12} = 563.$$

It is obvious that we need not have worked with all the points shown in Figure 10.7; because of symmetry, it would have been sufficient to work on points 6, 7, 12, only. However, at the beginning it is best to use the whole figure until one develops a feeling for the symmetry of the block operators.

The strength of the block-operator approach is even more evident when the region is larger and the number of points greater.

```
        16                    17                    18
  0  ●  500            0  ●  500            0  ●  500
344     688          94     625          344     688
        0           344     343                    0
                           -1

        11                    12                    13
  0  ●  500            0  ●  500            0  ●  500
 94     625          125       0           94     625
344     343           94                  344     343
       -1             94                          -1
                     344

         6                     7                     8
  0  ●  500            0  ●  500            0  ●  500
344     688          94     625          344     688
        0           344     343                    0
                           -1
```

Fig. 10.7.

EXERCISES

1. Work through Example 10.6 with a mesh width of $h = .5$ (the work can be greatly reduced by using the approximate solution obtained in Example 10.6 as the initial trial solution in the new problem).

2. Apply relaxation to the problem

$$\frac{\partial^2 f}{\partial x^2} + \frac{\partial^2 f}{\partial y^2} = 0$$

in a square of side 5 with $f = -100$ along the boundaries.

3. Solve

$$\frac{\partial^2 f}{\partial x^2} + \frac{\partial^2 f}{\partial y^2} = -500$$

if the region of definition for f is

(a) the square bounded by $x = 0$, $x = 5$, $y = 0$, $y = 5$;
(b) the rectangle bounded by $x = 0$, $x = 6$, $y = 0$, $y = 12$;
(c) the triangle bounded by $x = 0$, $y = 0$, $x + y = 6$.

In each case f is zero along the boundaries of the regions.

4. Rework 3(a) if the equation is

$$\frac{\partial^2 f}{\partial x^2} + \frac{\partial^2 f}{\partial y^2} = -200x.$$

5. Given

$$\frac{\partial^2 f}{\partial x^2} + \frac{\partial^2 f}{\partial y^2} = -500,$$

with $f = -100$ along the lines $x = 0$, $x = 3$, $y = 0$, $y = 3$. Solve the equation by taking $h = 1$, and use the result obtained as a starting value to give a better solution by taking $h = .5$.

6. Solve the equation

$$\frac{\partial^2 f}{\partial x^2} + \frac{\partial^2 f}{\partial y^2} = -10(x^2 + y^2 + 10)$$

over the square with sides $x = 0$, $y = 0$, $x = 8$, $y = 8$, with $f = 0$ along the boundaries.

CHAPTER ELEVEN

The Principles of Automatic Computation

1. HAND AND MACHINE CALCULATIONS

A recent science-fiction novelette was based on the impact of the discovery of our present system of hand-multiplication on a future world which had forgotten it, a world in which every scientist, army officer, etc., carried around his own private pocket-computer. The idea of the novelette is admittedly extreme, but it does point up a very marked present-day trend—the increase of automation in obtaining answers to any mathematical problems involving actual numerical work. This development has led from mechanical methods of computation through hand-operated and electrically-operated desk calculators to the enormous complexities of the most elaborate electronic computers.

Probably the earliest calculator, and one which a skilled operator can still employ to great advantage, is the famous Japanese abacus. This instrument is available in curiosity stores, and many people are familiar with the story about the contest in which a Grade One abacus-operator more than held his own against a desk calculator. Actually, the abacus, in a modified form, is a common plaything for children in this country; most children recall having owned or come into contact with a set of beads strung

242

on parallel wires set in a wooden frame. These "children's beads" can be used as an abacus; for example, to add five and four, we place five beads on the first wire, then join four more with them, and thus obtain nine as the empirical answer; similarly, to add seven and eight, we start with seven beads, add three (this completes the first wire; so we carry one bead to the second wire and start over again on the first), and then add five more to give the answer fifteen (one bead on the second wire, five on the first). This very simple principle carries over into desk calculators. For the rest of this chapter, it must be presumed that the student has an acquaintance with the operation of desk calculators (which are really just younger brothers of electronic computers).

One of the simplest manually-operated calculators is the Monroe Educator. One can think of it as basically being a rotating drum with cogs on it (quite analogous to a set of wires with beads on them). Addition is performed just as on an abacus—seven cogs plus eight cogs yields five cogs in one position and one cog in the next higher position; subtraction is the reverse of addition. Multiplication is performed by continued addition; thus 7×3715 is found by adding 3715 a total of seven times (turning the crank on the machine a total of seven times). Multiplication by a larger number involves shifting so as to correctly position the multiplier; for example,

$$257(3715) = 7(3715) + 50(3715) + 200(3715);$$

note that only a single digit multiplies at any time, the zeros being obtained by shifting the carriage of the machine. The result is just like ordinary multiplication;

$$
\begin{array}{ll}
3715 & \\
\underline{257} & \\
26005 & \text{multiply by 7; shift left.} \\
18575 & \text{multiply by 5; shift left.} \\
\underline{7430} & \text{multiply by 2.} \\
954755 & \text{accumulate.}
\end{array}
$$

Division is similarly performed as a process of continued subtraction. Each operation done by hand has its exact analogue on the Educator (other simple models which clearly illustrate the identity of hand and manual-calculator techniques are the Brunsviga and the Walther).

The only way in which an electrically-operated calculator differs from a manually-operated one is in speed; instead of performing multiplication by turning a crank and shifting by hand, we press a button. The machine does its own turning and shifting and this speeds matters up enormously. But the principles involved are exactly the same; multiplication is still done by a process of successive additions, division by a process of successive subtractions. Because of its speed, the electrical desk calculator can handle problems that are rather cumbersome using only hand techniques. The iterative solution giving the largest root of

$$x^6 - 27x^5 + 105x^4 - 140x^3 + 81x^2 - 21x + 2 = 0$$

as 22.625853 (Example 4.9) is easily carried out on a desk calculator; by hand, the time involved is considerable. There are, however, many problems in modern science and engineering which require so much computation that they have, until recently, been insoluble; even using the best desk calculators, with dozens of operators, decades would be required to solve some of these problems. The advent of electronic computers has changed this picture completely; problems that would previously have required several men working for a century (an arrangement which would be impossible psychologically, and almost bound to be inaccurate) can now be done very quickly on an electronic computer. To mention a trivial example, a system of forty linear equations in forty unknowns is a monstrous problem for a human being; for a large electronic computer, it is quite a routine piece of work (provided that the equations are not ill-conditioned).

At this point, it might be in order to point out that the electronic computers discussed in this chapter are all electronic digital computers, which actually work with numbers. We are not concerned here with analogue computers, which obtain their results by building up electronic or mechanical circuits in which lengths, angles, voltages, or other continuous physical variables are used to represent the variables of a given mathematical problem, and which thus produce an answer, obtained physically and graphically, as the answer to the mathematical problem which has been simulated. Further, it must be stressed that this chapter does not, and could not, purport to be a deep treatment of electronic computers; a whole book would be needed for such a treatment, and it would necessarily involve a great deal of incomprehensible specialized jargon. What we want to do is to discuss some of the main ideas which are common to all electronic computers; each individual computer will have a host of very special features, but there are certain broad principles which do not depend on the particular brand of computer. It is these broad principles which we wish to sketch.

2. FUNDAMENTAL STRUCTURE OF AN ELECTRONIC COMPUTER

As we indicated in the last section, electronic computers handle problems on a large scale; some people even like to refer to them as "giant brains", a practice which should be avoided because of the philosophical nonsense to which it may lead.† One prominent mathematician has even been led into claiming that he has "proved" the finite span of human life from this analogy. His argument (paraphrased) runs as follows: computers are like brains, brains are like computers; the human brain has billions of memory cells—when these are full, you can no longer receive sensations—hence you must be dead! Thus your life-span is determined by the time it takes you to fill your memory bank. This argument was propounded quite seriously, although its proponent did not claim to have proved that his finite life-span was "three score years and ten".

At this point, the following quotation from E. T. Bell (*Mathematics, Queen and Servant of Science*, p. 4) is pertinent. "In the nineteenth century, 'the mind' was mechanically dissipated in the crude steam-engine and energy analogies then fashionable. The echo of all that furious nonsense has bounded back, amplified but recognizable, in the assertion that the human nervous system, including the 'thinking brain', mimics an electronic supergadget, not the other way about."

Putting aside the brain concept, and admitting that computers are nothing more than extremely elaborate machines, let us look at the fundamental parts of these machines. These can be summarized as follows: an input mechanism, whereby data and instructions may be fed into the machine; an operating unit which does the actual numerical work; a memory unit which stores facts for use during later parts of the analysis; an output mechanism which prints or records the results in some manner. The operating unit can be further subdivided into a control unit, which interprets the instructions received from the input, and an arithmetic unit, which does the required calculations. The memory unit can likewise be subdivided into a working memory, for immediately required facts, and a reserve memory, for more remotely required facts. Using the dubious human analogy, the input unit is the machine's eyes (reading the instructions); the control unit is the brain (interpreting instructions); the arithmetic unit acts as hands and pencil (performing calculations); the memory unit is the memory; the output unit is a second pair of hands (used for writing down results).

† A great deal of interest has currently been evoked by attempts at machine-translation from one language to another. The resulting curious mixtures can sometimes even be understood. It is likely that some day machines can be used to translate technical articles, where the words involved tend to be concrete, and limited in the number of meanings which may be attached to them; even then, the results will likely be utilitarian, ungrammatical, and awkward. Literary and figurative language will always remain beyond the scope of machine-translation. Matthew Arnold has already remarked that, in the case of Shakespeare, "Canst thou not minister to a mind diseased?" is not well paraphrased by "Can you not wait upon a lunatic?", and a machine could hardly attain even this plebeian and prosaic level. Rather, a double translation from English to Russian to English can change "The spirit is willing but the flesh is weak" into "The vodka is strong but the meat is rotten".

In later sections we shall discuss input and output mechanisms briefly; at the moment, let us consider the operating unit and the memory unit. They have in common the property that both must deal with numbers (either in use, or waiting to be used); so they must have a method of recording numbers. Such recording is usually achieved by the use of magnetic drums or magnetic cores. On a magnetic drum there are thousands of spots which may either be magnetized in one direction or the other; if a spot is magnetized in one direction, we say that it records a number zero; if it possesses opposite polarity, we say that it records a number one. Thus numbers can readily be recorded on a magnetic drum if they can be built up out of zeros and ones; this is achieved by employing the binary number system, which will be described in the next section.

In concluding this section, we might say a few words about other systems of storing and recording data. Early computers used cathode-ray tubes (electrostatic storage of this nature has the disadvantage that you lose the data when the power is turned off) or mercury delay lines. Some machines still use condenser storage either partially or completely; such storage is inexpensive, but is a disappearing storage—it is lost when the power is turned off. A variation on magnetic drum storage has been made by the use of magnetic disks (having a drum built up out of closely placed disks greatly increases the surface area available for storage of numbers). Finally, doughnut-shaped magnetic cores are superior to drums in that they are faster and more reliable; the data are immediately available since there is no time lag, as on a drum, where one must wait for the drum to rotate to a point where the reading heads may pick off the required data.

3. THE BINARY NUMBER SYSTEM

When we write down an ordinary number, such as 8294, we are really recording this number in terms of powers of ten; thus

$$8294 = 8(10)^3 + 2(10)^2 + 9(10) + 4,$$

that is, the digits appearing in the number are the coefficients of successive powers of ten in its ordinary decimal expansion. Similarly,

$$375.624 = 3(10)^2 + 7(10) + 5 + 6(10)^{-1} + 2(10)^{-2} + 4(10)^{-3}.$$

A non-terminating decimal is really an infinite series; for example,

$$423.14159265\ldots = 4(10)^2 + 2(10) + 3 + 1(10)^{-1} + 4(10)^{-2} + 1(10)^{-3} + \ldots.$$

For convenience, these decimal expansions, using powers of ten, are always used in everyday life (although a few dedicated souls keep advocating a change to a duodecimal system using powers of twelve); however, there is clearly no mathematical reason for using ten (beyond the outstanding convenience of the choice), and we might write

$$8294 = 7(1184) + 6 = 7[7(169) + 1] + 6$$
$$= 7^2(169) + 7(1) + 6$$
$$= 7^2[7(24) + 1] + 7(1) + 6$$
$$= 7^3[7(3) + 3] + 7^2(1) + 7(1) + 6$$
$$= 3(7)^4 + 3(7)^3 + 1(7)^2 + 1(7) + 6.$$

We should then say that the ordinary number 8294 had the expression 33116 *in the scale of seven.* Clearly this process is quite general, and we can formulate the

DEFINITION. If a number x is expressed in the form

$$x = a_k r^k + a_{k-1} r^{k-1} + \ldots + a_1 r + a_0 + a_{-1} r^{-1} + a_{-2} r^{-2} + \ldots ,$$

then we write

$$x = a_k a_{k-1} \ldots a_1 a_0 . a_{-1} a_{-2} a_{-3} \ldots ,$$

and say that x is expressed in the scale of r. The number r is called the radix or base of the scale of notation. Obviously $0 \le a_i < r$ for all i, that is, at most r distinct digits appear in the expansion of x in the scale of r.

EXAMPLE 11.1. Express 954.4 in the scale of three.
It is readily verified by division that

$$954 = 1(3)^6 + 0(3)^5 + 2(3)^4 + 2(3)^3 + 1(3)^2 + 0(3) + 0;$$

the successive divisions required are usually recorded in the following set-up.

$$
\begin{array}{r}
3 \,\lfloor 954 \\
3 \,\lfloor 318 + 0 \\
3 \,\lfloor 106 + 0 \\
3 \,\lfloor 35 + 1 \\
3 \,\lfloor 11 + 2 \\
3 \,\lfloor 3 + 2 \\
1 + 0
\end{array}
$$

Thus the integral part of the given number is 1022100, in the scale of three.
On the other hand, successive multiplications by 3 give

$$\frac{4}{10} = \frac{12}{30} = \frac{1}{3} + \frac{2}{30},$$
$$\frac{2}{30} = \frac{6}{90} = \frac{0}{9} + \frac{6}{90},$$
$$\frac{6}{90} = \frac{18}{270} = \frac{1}{27} + \frac{8}{270},$$
$$\frac{8}{270} = \frac{24}{810} = \frac{2}{81} + \frac{4}{810},$$
$$\frac{4}{810} = \frac{12}{2430} = \frac{1}{243} + \frac{2}{2430},$$
$$\frac{2}{2430} = \frac{6}{7290} = \frac{0}{729} + \frac{6}{7290},$$
$$\cdot \quad \cdot \quad \cdot \quad \cdot \quad \cdot \quad \cdot \quad \cdot \quad \cdot \quad \cdot \quad \cdot \quad \cdot \quad \cdot \quad \cdot \quad \cdot$$

The process is now repeating and, collecting terms, we find

$$954.4 \text{ (scale 10)} = 1022100 \,.\, 10121012\dot{1}01\dot{2} \dots \text{ (scale 3)}.$$

EXAMPLE 11.2. Change $5!2.3\dot{5}$ (scale 7) into an ordinary decimal.
The given number represents

$$5(7)^2 + 1(7) + 2 + \frac{3}{7} + \frac{5}{7^2} + \frac{5}{7^3} + \frac{5}{7^4} + \cdots$$

$$= 254 + \frac{3}{7} + \frac{5}{42} = 254\frac{23}{42} = 254.547619 \dots \text{ (scale 10)}.$$

EXAMPLE 11.3. How many digits are required to express 10^6 in the scale of two?
Suppose that

$$2^x < 10^6 < 2^{x+1}.$$

Then

$$10^6 = a_x 2^x + a_{x-1} 2^{x-1} + \dots + a_1 2 + a_0,$$

and we see that $x + 1$ digits are required. We can express this result in symbols as $1 + [\log_2 10^6]$, that is, $1+$ the greatest integer in $\log_2 10^6$.

The scale of notation with $r = 2$ is called the *binary* scale; computations by electronic computers are carried out with numbers expressed in binary notation. The reason is that, while any number x requires more digits to express it in the binary scale than it does in any other scale, these digits are very simple, being either zeros or ones. Thus the binary scale lends itself admirably to recording on a magnetic drum by the "Yes–No" method of the direction of magnetization of a spot; one direction of magnetization is interpreted as a zero, while the opposite direction denotes a one. The number 38 (100110 in the binary scale) can thus be recorded as *PNNPPN*, where *P* and *N* denote the two directions of magnetization.

EXAMPLE 11.4. Change 1947 to binary notation.

$$\begin{aligned}
1947 &= 1024 + 923 \\
&= 1024 + 512 + 411 \\
&= 1024 + 512 + 256 + 155 \\
&= 1024 + 512 + 256 + 128 + 27 \\
&= 1024 + 512 + 256 + 128 + 16 + 8 + 2 + 1 \\
&= 2^{10} + 2^9 + 2^8 + 2^7 + 2^4 + 2^3 + 2 + 1 \\
&= 11110011011 \text{ (scale 2)}.
\end{aligned}$$

EXAMPLE 11.5. Change 110101010 (scale 2) to scale 10.
The number represents

$$2^8 + 2^7 + 2^5 + 2^3 + 2 = 426 \text{ (scale 10)}.$$

EXAMPLE 11.6. If $a = 111010$, $b = 1011$ (scale 2), evaluate $a + b$, $a - b$, ab, a/b.

$$
\begin{array}{r}
111010 \\
1011 \\
\hline
a + b = 1000101
\end{array}
\qquad
\begin{array}{r}
111010 \\
1011 \\
\hline
a - b = 101111
\end{array}
$$

$$
\begin{array}{r}
111010 \\
1011 \\
\hline
111010 \\
111010 \\
000000 \\
111010 \\
\hline
ab = 1001111110
\end{array}
$$

$$
a/b = 101.01000101 \ldots
$$

$$
\begin{array}{r}
1011\,|\,111010 \\
1011 \\
\hline
1110 \\
1011 \\
\hline
1100 \\
1011 \\
\hline
10000 \\
1011 \\
\hline
10100 \\
1011 \\
\hline
1001 \text{ etc.}
\end{array}
$$

Further experience is provided in the

EXERCISES

1. Convert the following numbers to scale ten:

    352 (scale 6),   714.12 (scale 8),   44.42 (scale 5).

2. Why can the number 346201 not be in scale 5?

3. The digits t (ten) and e (eleven) occur in scale 12; convert $5t2e$ (scale 12) to scale 10; convert 1699 (scale 10) to scale 12.

4. Check Example 11.6 by performing the operations in scale 10.

5. Convert 1111101 (scale 2) to scale 5; convert 6543 (scale 7) to scale 2.

6. Amplify the following algorithm to show why it expresses $\frac{4}{9}$ as the repeating binary decimal .$\dot{0}1110\dot{0}$.

$$
\begin{array}{ccc}
 & & 4 \\
8 & 0 & 8 \\
16 & 1 & 7 \\
14 & 1 & 5 \\
10 & 1 & 1 \\
2 & 0 & 2 \\
4 & 0 & 4 \\
8 & 0 & 8 \\
16 & 1 & 7 \\
\end{array}
$$

. . . .

(The first entry in any row is always found by doubling the last entry in the preceding row; the second and third entries in any row are the quotient and remainder on dividing the first entry by 9.)

7. Use an algorithm similar to that in Exercise 6 to express .26 (scale 10) in scale 3 (the ternary scale).

8. Work out the following expressions in the binary scale.

   (a) $101 + 11 + 110110 + 110101 - 1101 - 1010$
   (b) $1111 - 111010 + 110111 - 110 + 100100$
   (c) $111^2 - 110^2$
   (d) $1101^3 - 101^3$
   (e) $(11011)(101101)$
   (f) $101111001/10111$
   (g) $(101.1011)(.1101)$
   (h) $11.01 - 1.0101 + (.011)(101.11)$

4. THE MIND OF THE MACHINE

The title of this section has been selected because it is a contradiction in terms; a computing machine has no mind. So, in answer to the question "How does the machine think?", we simply reply "It does not think at all; it merely operates". And the operation is conditioned by one basic principle: a computer can only obey orders; it will only do what you have instructed it to do.

The fact that no exercise of judgment is possible to the machine predetermines some of the types of orders that it must receive. For example, it does not know that $\sqrt{-5}$ can not be extracted as a real number; so, if in the course of a problem, it is instructed to find $\sqrt{a}$, where a is recorded in Register Number 7, and if Register Number 7 contains -5, difficulties arise; the machine may give a very weird answer. The people who like to personify machines say that in a case like this, confronted by an impossible task, the machine goes psychotic. However one cares to phrase it, the fact remains that the "giant brain" must be told what to do; it should never be given an instruction requiring it to "find the square root of a, where a is recorded in Register 7". Rather, the instruction should be of the nature "Let a denote the contents of Register 7; if a is positive, find $\sqrt{a}$; if a is negative, find $\sqrt{-a}$ and print the symbol i after the number thus obtained". Similarly, any set of instructions given to the machine should forestall any possible divisions by zero.

Since even the tiniest details of procedure must be contained in the instructions given to the machine, any problem must be completely analysed by a human before a programme of instructions is prepared for the machine. The programme will be most advantageous if it contains very simple procedures. The reason for this is that almost any procedure used must be broken down into the basic operations of addition, subtraction, multiplication, and division; any complicated procedure requires inclusion of an entire sub-programme to handle it. Consequently, one prefers programmes where the steps required stay as close as possible to the basic operations. Because of the fact that most computers (like desk calculators) perform multiplication by repeated additions and division by repeated subtractions, multiplication and division take appreciably longer times than do addition and subtraction. (For example, one common computer will do an addition or a subtraction in about three-quarters of a millisecond; a multiplication with a five-digit multiplier may take 6 or 7 milliseconds, and a division with a five-digit quotient may take 9 or 10 milliseconds.) Other operations, such as extracting square roots, which must, in general, be programmed as composite operations, take much longer. So if a problem can be worked in two ways, and if the first way involves considerable mathematical complication, whereas the second is long, cumbersome, inelegant, but involves only very simple mathematical operations (usually repeated a vast number of times), then it can probably safely be stated that the second method is superior for use on the computer. The complicated method may be preferred for hand computation, where every step requires time, and where the number of steps must hence be limited; but the computer can work so speedily that time is relatively unimportant— what is important in preparing a programme for the computer is to have as simple and direct a set of instructions as possible—complications should be avoided (cf. Example 11.8).

As an example of a process which can encounter difficulties when one tries to use it with an electronic computer, we mention the relaxation procedure. Here, at any stage, one has to decide what residual to tackle next; this requires an act of judgment. Use of some general rule such as "Always liquidate the largest residual remaining at any stage" would allow one to build up a machine programme, although not necessarily an economical one; but such a general rule would probably be far from the best possible procedure (and one can hardly ask a machine to use its non-existent feminine intuition in selecting its successive steps). This difficulty points up the fact that processes are best for an electronic computer if they are routine; processes requiring ingenuity, variation, and judgment, such as the relaxation process, are troublesome. On the other hand, the Gauss-Seidel or Jacobi methods of iteration are good machine procedures; once started, the steps follow one another in a rigid pattern; so the machine can receive perfectly definite orders as to what it is to do at any stage.

Indeed, an electronic computer is especially good for such repetitive problems; among the types of problems in which the computer does shine are cases where

(a) the same type of problem is to be solved a great number of times (that is, the problem remains constant, but the parameters change); in such

Table 11.1

							x
29046.816	-58.979592	-91.428572	-39.102041	-23.693877	8.6530612	11.714286	-20.306122
-58.979592	15.525510	2.9642857	3.8724490	4.1326531	-.43367347	1.3928571	3.6173469
-91.428572	2.9642857	28.750000	4.6785714	1.2142857	2.1071429	-.25000000	9.0357143
-39.102041	3.8724490	4.6785714	12.137755	5.3367347	1.6683673	.53571429	3.4132653
-23.693877	4.1326531	1.2142857	5.3367347	9.4897959	.74489796	.64285714	3.0102041
8.6530612	-.43367347	2.1071429	1.6683673	.74489796	4.8724490	-.17857143	1.0051020
11.714286	1.3928571	-.25000000	.53571429	.64285714	-.17857143	6.7500000	1.6071429
-20.306122	3.6173469	9.0357143	3.4132653	3.0102041	1.0051020	1.6071429	

Table 11.2

3.5114740	10.272168	10.514395	5.7351173	1.8446722	-12.019436	-8.3115732	-1.9381797
10.272168	7798.9031	-484.22230	-1225.2481	-2558.3592	1729.6981	-1169.9761	-371.39750
10.514396	-484.22230	4178.4083	-1346.0341	867.00601	-1303.1747	479.74998	-1064.8755
5.7351169	-1225.2481	-1346.0342	12168.576	-5845.5940	-2793.1572	-259.39895	-128.13341
1.8446723	-2558.3591	867.00600	-5845.5945	15145.158	-777.86097	-267.15882	-735.22730
-12.019435	1729.6980	-1303.1747	-2793.1571	-777.86104	22399.241	551.64488	-200.08648
-8.3115728	-1169.9760	479.74990	-259.39899	-267.15896	551.64486	15330.252	-761.02340
-1.9381797	-371.39749	-1064.8755	-128.13340	-735.22730	-200.08649	-761.02353	3681.2260

a case the programme is probably available among the library routines maintained for the particular machine;

(b) the solution is obtained by an iteration, that is, the answer at any stage is obtained by substituting into some expression the answer from the previous stage;

(c) the answer results from straightforward application of an orderly sequence of steps, usually similar, and often repeated (this occurs in matrix inversion, and, of course, any iteration can itself be considered as an orderly sequence of similar steps).

We might conclude this section with two examples of problems which illustrate the tremendous advantage of computers over desk calculators in speedy consummation of a large number of elementary operations.

EXAMPLE 11.7. Invert the matrix A and solve the system of equations $AX = B$; A is the 8×8 matrix given in Table 11.1, and B is the column vector with components .63166667, .15666667, .34416667, .13541667, .10416667, .05208333, .05250000, .28375000.

We have omitted one entry in A for problem purposes. Also, we should note that in this actual problem A^{-1} was required; naturally, it would be less work to find X without determining A^{-1}.

The computer which actually handled this problem uses essentially the procedure of Example 7.11, that is, it reduces A to the identity I and then performs the same operations on I to yield A^{-1}. Thus it carries through a large number of divisions, multiplications, additions, and subtractions. The result is obtained in a couple of minutes operating time, and is recorded in Table 11.2 (the matrix given in Table 11.2 is $10^5 A^{-1}$). Using Table 11.2, $10^5 X = 10^5 A^{-1} B$ is found to be a vector with components 6.8025476, 552.52943, 932.04140, 191.86578, 421.61986, 494.43109, 531.24092, 474.33087.

EXAMPLE 11.8. The equation

$$(\cot nr - \coth nr)(\sin mr \cosh mr - \cos mr \sinh mr)$$
$$+ \, 2 + 2 \cos mr \cosh mr = 0$$

occurred in an engineering research problem. Here r is the unknown, and m and n are parameters with the property that $m + n = 1$. It is required to solve the equation as the parameter m assumes values from $0(.1)1$.

Here the problem is quite complex in form, and the Bolzano procedure (Example 4.11) was actually used. An interval containing a root was located, and then successive bisections were employed to keep halving the interval. Twenty bisections were actually sufficient for the requirements of the problem.

It is worth noting how very simple the Bolzano process is in this example. The Bolzano process only requires time to keep narrowing down the interval.

This time is no problem for an electronic computer, because it can perform such a vast number of operations while a human being would only be doing one.

EXERCISES

1. In Example 11.7, obtain eight equations for the missing entry x by multiplying the last row of A by the successive columns of A^{-1}. Show that the best estimate for x is 31.239793.

2. In Example 11.7, compute the results of (a) multiplying the first row of A by the first column of A^{-1}, (b) multiplying the second row of A by the fifth column of A^{-1}.

3. In Example 11.7, compute the results of (a) multiplying the last row of A by the column vector X, (b) multiplying the first row of A^{-1} by the column vector B. In each case, check the results by comparing with the theoretical values. What causes the slight discrepancies?

4. Solve the equations $AX = B$, where A is given in Example 11.7 and B is the vector with components .12, .45, .37, $-.25$, $-.87$, .65, .04, $-.97$.

5. Repeat Exercise 4 if X has components $-.27$, .88, $-.09$, $-.30$, .76, .43, $-.12$, .24.

6. Suppose that a machine follows the following rule: the number of milliseconds for an addition is .096, for a multiplication is $.096(20 + 2d_1)$, for a division is $.096(40 + 2d_2)$, where d_1 is the sum of digits in the multiplier, and d_2 is the sum of the digits in the quotient. Compute the time required for
   (a) multiplication by 394,
   (b) multiplication by 1942576,
   (c) division with a quotient of 67294,
   (d) division with a quotient of 3124,
   (e) division with a quotient of 95402591.

5. METHODS OF PROGRAMMING

We have already described how numbers actually used within a computer can be coded into binary notation. Originally, all numbers were fed into computers in binary notation; however, this took time, and now many machines have input mechanisms whereby orders are given to the machine using ordinary decimal notation (and ordinary operation signs, such as $+$ and $\times$). The machine inwardly translates into binary notation, carries through the computations in the binary system (or in a binary coded decimal system), translates, and then prints the result in the ordinary scale of 10.

EXAMPLE 11.9. A small computer was used to find the largest root of the equation

$$x^7 - 27x^6 + 15x^5 - 21x^4 - 80x^3 - 32x^2 - 17x - 3 = 0.$$

Using the method of Example 5.9, we write

$$x = 3\,]\frac{1}{x} + 17\,]\frac{1}{x} + 32\,]\frac{1}{x} + 80\,]\frac{1}{x} + 21\,]\frac{1}{x} - 15\,]\frac{1}{x} + 27.$$

The particular machine used on this very simple problem is not elaborate; orders are given by simply pressing typewriter keys; the keys punch a paper tape which feeds into the machine. The answer is then recorded by an electric typewriter attached to the machine. The programme printed in Table 11.3 gave the machine its orders.

The successive convergents were printed out as follows.

27.+	26.477 37685 60289+
26.477 37685 60289+	26.467 80993 16370+
26.467 80993 16370+	26.467 63159 00404+
26.467 63159 00404+	26.467 62826 43731+
26.467 62826 43731+	26.467 62820 23565+
26.467 62820 23565+	26.467 62820 12000+
26.467 62820 12000+	26.467 62820 11785+
26.467 62820 11785+	26.467 62820 11781+
26.467 62820 11781+	26.467 62820 11780+
26.467 62820 11780+	26.467 62820 11780+

The matrix inversion performed in Example 11.7 can serve to illustrate another common method of programming. There the data (matrix elements) were recorded on punched cards; these were fed into the machine; the set of orders necessary to carry out the many operations of the method of Example 7.11 was given; the answer was recorded on punched cards. From this brief summary, we can see the importance of several points.

(a) It is very advantageous to use "off-line equipment". The small computer mentioned in Example 11.9 uses "machine time" to read in the instructions. However, for a large computer, the computer's time (and at the present-day rents for computers, this is an important economic consideration) is saved if the original data are punched on cards or on paper tape by auxiliary equipment. Similarly, it is desirable to have auxiliary equipment to print the results from the output cards.

(b) Whether the machine receives its orders from paper tapes, punched cards, wired boards, or other methods, it is important, to build up a library of programmes commonly used. For example, if punched cards are being used, it would be intolerable to have to write a programme every time one wanted to invert a matrix. Instead, one

Table 11.3. PROGRAMMING OF EXAMPLE 11.9
(slightly modified)

Register	Operation	Operand	Explanation
00	Enter	27.0+	During these steps, the constant data are fed
01	Enter	3.0+	into the machine; this is done merely by
02	Enter	17.0+	pressing typewriter keys. For example, the
03	Enter	32.0+	first line is achieved by pressing keys
04	Enter	80.0+	labelled 0 0 ENT 2 7 . 0 +
05	Enter	21.0+	
06	Enter	15.0−	
07	Enter	27.0+	

Punch	On		Switch settings for the machine are made.
Automatic Decimal Point Mode On			
Punch Class	Zero		

Register	Operation	Operand	Explanation
01	Divide	00	The answer is being obtained in the Answer
	Add	02	Register (the A Register); these instruc-
	Divide	00	tions are given by depressing the appro-
	Add	03	priate typewriter keys in the input
	Divide	00	mechanism; for instance, the first keys
	Add	04	depressed will be
	Divide	00	0 1 ÷ 0 0 + 0 2 ÷ 0 0
	Add	05	
	Divide	00	
	Add	06	
	Divide	00	
	Add	07	

Register	Operation	Operand	Explanation
00	Copy	A	The answer has been obtained in Register
	CR		A; this answer is now moved to Register
	SL15		00; the carriage is returned to position
	RO		(that is, the answer typewriter is turned on and the carriage is positioned at the left-hand margin); the left-hand zeros on the answer are removed; the answer (now in Register 00) is read out on the answer typewriter.

Punch	Off		The programme tape continuously re-creates itself, and the same programme is
Duplicator	On		followed again and again. Note that the
Programme Tape Reader On			first value entered in Register 00 was 27; however, the result of the first iteration is now in Register 00, and it now serves as the beginning of the second iteration. The process continues.

keeps the "matric-inversion programme " on file; then it is fed to the machine whenever one needs to invert any particular matrix. In time, an impressive library of standard programmes can be assembled for any machine.

(c) Machine time is saved if "buffering" is introduced; this is simply the procedure whereby the operational unit of the machine is simultaneously receiving data, computing, and recording results. More time would be consumed if all the data had to be fed in before computation started, and if all computation had to cease before recording of results could begin. The idea of buffering actually appears in some desk calculators such as the Marchant. If one multiplies a by b, the computation in the calculator and the recording in the upper register are both going on while one is still punching in the multiplier b. A somewhat similar principle of simultaneous computation, on a more grandiose scale, constitutes buffering in an electric computer. (Naturally, the analogy with a desk calculator is only approximate.) Buffering is important in helping to avoid bottlenecks at the input and output stages.

(d) The magnetic drum is limited in the amount of information it can hold; it serves as a working memory to hold instructions and data which are just about to be used. However, the drum is not large enough to be used as a reserve memory. Suppose, for example, that in some problems we needed a list of all primes from 1 to 10,000; it would be wasteful to store such seldom-used data on the working drum and thus inutilize space that one might need. Instead, such data could be recorded on a magnetic tape, and this tape fed into the machine for those problems (and only those problems) where it was needed. A further use for magnetic tapes is as a subsidiary memory in very long and complex problems; here one may encounter "partial answers", that is, results which are not end-results but which will be needed later in the problem. However, if there are many partial answers, the working memory may not suffice to hold all of them; in this case, the partial answers can be stored temporarily on magnetic tapes, brought up and used when they are needed, and then erased.

(e) Magnetic tapes provide a faster input than either punched tape or cards. A machine can read instructions from tapes up to 250 times as fast as from punched cards.

(f) A stored-programme computer (that is, one with programme stored on the drum) is the most flexible type; a computer with a stored programme can not only manipulate numbers, but can also manipulate its own instructions; this ability to change the steps in the instructions is not present if the programme is supplied by punched paper tape, a wired panel, or a plugboard with a series of linking pins.

General principles, of the nature of these mentioned under (a)–(f), are featured by almost all electronic computers.

6. FINAL WORDS

In this section, we collect four general remarks which should be made in conclusion.

The first remark is that the two commonest procedures (and ones which should, therefore, be very simple to programme) in any numerical computation are:

(a) let p be the previous result; form $p + ab$, where a and b are given constants (this type of order crops up again and again in Examples 7.11 and 11.7 on matric inversion);

(b) let p be the previous result; form $(p + a)b$, where a and b are given constants (this type of order is well exemplified by Example 11.9).

Our second remark concerns errors. A desk calculator may occasionally make inexplicable errors; if a calculator has made an error using one routine, it is liable to repeat the error if the same routine is repeated. With desk calculators, the best check is to do the problem a second time on another machine (preferably using a different method); if this can not be done, a repeat on the original machine, using a different method, is fairly reliable. Many electronic computers have built-in checks which serve to detect most errors; other computers rely almost exclusively on programmed checks. A computer is unlikely to make the same random error twice; inaccuracies are usually due not to the electronic circuits but to deficiencies in the mechanical input or output (such as failure to punch a hole in a tape).

We now answer a question which must for some time have been gnawing at the student's mind. We have talked about general principles involved in programming for and in computation by electronic computers. But we have not given specific instructions on how to work on a computer. The reasons for this are two. First, it would take a great deal of space (see, for example, the book on programming instructions which accompanies any electronic computer). Secondly, it would be futile, because all computers (like all women) are very different. Experience in the *details* of an instruction *code* for machine A is very good practice in the instruction code for machine A; but it is not good practice in instruction-coding for machine B. Each type of computer is so distinctively specialized that it must be studied *per se*. Naturally, general experience gained in working with one computer is of great value in working with another computer; the same broad procedures for writing *programmes* will crop up with both computers; but experience concerning *specific coding features* of machine A is pretty well confined, in applicability, to A itself. Consequently, the student is best advised to study details of instruction-coding in connection with whatever machine is available to him. With the increased use of electronic computers in universities and in industry, this should be feasible either during university course-work or during practical experience in summer employment.

Finally, we conclude with a word of warning. The student should not get the idea that electronic computers have removed the need for desk calculation and hand calculation. There is a natural progression in techniques, and this is determined by the time required for a problem and the complexity of the problem. Thus, one normally multiplies 7 by 12 mentally, not on paper; and one normally multiplies 157 by 346 by hand, not on a desk calculator (even if available). Our progression then is: mental calculation, hand calculation, desk calculation, computation by an electronic computer. A problem must possess a certain degree of complexity before it is worth-while solving it on an electronic computer (from the points of view of both time and money). So there is no possibility that manual calculation and desk calculation will become outmoded if for no other reason than the one we have previously stressed: that every programme written for a computer requires a detailed analysis of the problem by a human operator before it ever reaches the computer. Indeed, in order to ensure that a programme prepared for a computer contains no errors or "bugs", the best approach is to take a simple problem; obtain the answer to it by following the programme on a desk calculator; and then see that the programme, when used by the computer on the same problem, produces the known answer obtained on the desk calculator.

Selected Bibliography

The student will find the following references of use. It is hoped that, having been supplied with such a brief list, he will not be discouraged from consulting some of the works!

GENERAL TEXTS:

D. R. Hartree, *Numerical Analysis*, Oxford, 1952.
A. D. Booth, *Numerical Methods*, Butterworths, 1955.

FINITE DIFFERENCES:

H. Freeman, *Mathematics for Actuarial Students, Part II*, Cambridge, 1939.

DIFFERENTIAL EQUATIONS:

H. Levy and E. A. Baggott, *Numerical Solutions of Differential Equations*, Dover, 1950.
W. E. Milne, *Numerical Solution of Differential Equations*, Wiley, 1953.

RELAXATION METHODS:

F. S. Shaw, *Relaxation Methods*, Dover, 1953.

INDEX

Polynomial equations, 6
   Graeffe's method for solution, 84, 95
   Horner's method for solution, 84
Polynomial interpolation, 52, 53
Possible error, 4
Product matrix, 166

Q

Quadratic equation, solution by iteration,
   78, 79
Quadratic factors, iterative procedure for
   real, 98-104

R

Ratio, Cauchy, 107, 113
Regression, 56, 57
Regula falsi, 82, 83, 87, 95
Relaxation:
   method, 188-192
   technique, 191
Relaxation operator:
   block, 238
   line, 222
   point, 221
   wedge, 224
Relaxations, 189
Residual, 189
Rolle diagram, 51
Roots:
   complex, 95-104
   multiple, 89-93
   near-multiple, 93, 94
Round-off error, 3, 23, 61
   in series, 109
Runge-Kutta formula:
   order four, 152, 155
   order n, 152
   order three, 154
Runge-Kutta method, 151-155

S

Scale of notation, 247
   base of, 247
   radix of, 247

Series:
   absolutely convergent, 107
   alternating, 109, 110
   convergent, 106, 107
   definite, 106
   differentiation of, 112
   divergent, 106, 107
   Euler-Maclaurin, 128, 129, 135
   geometric, 107
   indefinite, 106
   infinite, 106 ff.
   integration of, 112
   insertion of parentheses, 107
   Lagrange, 137-139
   Maclaurin's, 112
   Maclaurin's bounds on value of, 108
   rearrangement of terms, 107
   round-off error, 109
   Taylor's, 112-116, 146, 147
   truncation error in, 109, 110
Shaw, F. S., 260
Sheppard's zigzag rule, 29, 30
Sigma notation, 5
Significant figure, 3
Simpson's rule, 116, 117
   closed, 122
   error in, 126, 127
   open, 122, 150
Singular matrix, 173, 176
Sokolnikoff, I. S., 113
Solution of equations:
   connection with inverse interpolation,
      68
   graphic, 68-72
   Newton's method for, 84-98
Square root, Newton's method for, 88
Staircase method, 81
Stirling's central difference formula, 37, 38
Stirling's formula for large factorials, 111,
      136, 137
Stokes' law, 77
Student's t, 61
Subscript notation, 10
Subtabulation, 21-23, 44, 45
Summation by parts, 212
Symmetric functions, elementary, 6
Symmetric matrix, 179